D0938039

PRAISE FOR

The Mystery of Metamorphosis

"Prepare your mind for its own Cambrian explosion. Frank Ryan's breath-taking explanations for life's most wondrous transformations are by turns shocking, thrilling, and revelatory."

—SY MONTGOMERY, award-winning naturalist and author of *Birdology* and *The Good, Good Pig*

"We tend to think of every detail of an animal's appearance, and most of its behavior, as the product of genes tweaked by environmental influences that induce slight changes in the outcome, as though that explains everything. So, can there be anything more mysterious than, say, a flesh-eating maggot turning a few days later into an iridescent green bottle fly on whom every single bristle is accounted for, or a leaf-eating tomato hornworm caterpillar becoming a hummingbird-like analogue hovering to suck nectar from flowers? Who are the biologists bold enough to tackle such a problem of profound transformation where 'environment' as we normally think of it has no bearing whatsoever? Has this not been THE most exciting story in biology? This deeply informed account delves into the fascinating history of the study of metamorphosis. It's a saga, and Frank Ryan does it justice."

—BERND HEINRICH, author of *The Nesting Season, Winter World,* and *Mind of the Raven*

"In this scintillating book Frank Ryan gives us a skillful and detailed introduction to the intricate world of animal metamorphosis and to the astounding idea (pioneered by Donald Williamson) that the larval and adult forms of many animals were once wildly different creatures whose genomes melded during moments of extraordinary trans-species hybridization. Read this book to experience the feel and excitement of a massive paradigm shift in science as told by a brilliant science writer."

—STEPHAN HARDING, author of *Animate Earth: Science, Intuition and Gaia.*

"Frank Ryan tells the story of animal metamorphosis by letting you follow the explorations of scientists who have researched the remarkable transformations of animal larvae into their adult form. . . . You not only learn about the rich and fascinating phenomenon of animal metamorphosis, you also learn about science as a complex human endeavor in which new ideas are hardly welcomed with open arms."

—CRAIG HOLDREGE, director, The Nature Institute

"An impressive work. Frank Ryan's thorough detection and crisp writing expertly bring forth the simple but profound discoveries that shape our understanding of metamorphosis, and challenge our ideas about evolution. The idea that dissimilar sea creatures can link together to create new creatures, and thus encourage evolution, is fascinating; and, among the scientists Ryan introduces, one makes a convincing case for just that sort of wonder."

—Brian Garfield, author of *Hopscotch* and *The Meinertzhagen Mystery: The Life and Legend of a Colossal Fraud*; past director, Mystery Writers of America

Also by Frank Ryan

The Forgotten Plague
Virus X
Darwin's Blind Spot
Virolution

THE MYSTERY OF METAMORPHOSIS

A Scientific Detective Story

FRANK RYAN

CHELSEA GREEN PUBLISHING
WHITE RIVER JUNCTION, VERMONT

Project Manager: Patricia Stone
Editorial Contact: Joni Praded
Developmental Editor: Jonathan Cobb
Copy Editor: Cannon Labrie
Proofreader: Susan Barnett
Indexer: Peggy Holloway
Designer: Peter Holm,
Sterling Hill Productions

Printed in the United States of America
First printing March, 2011
10 9 8 7 6 5 4 3 2 1 11 12 13 14 15

Chelsea Green is committed to preserving ancient forests and natural resources. We elected to print this title on 30% post-consumer recycled paper, processed chlorine-free. As a result, we have saved:

14 Trees (40' tall and 6-8" diameter)
4 Million BTUs of Total Energy
1,285 Pounds of Greenhouse Gases
6,188 Gallons of Wastewater
376 Pounds of Solid Waste

Chelsea Green made this paper choice because our printer, Thomson-Shore, Inc., is a member of Green Press Initiative, a nonprofit program dedicated to supporting authors, publishers, and suppliers in their efforts to reduce their use of fiber obtained from endangered forests.

For more information, visit www.greenpressinitiative.org

Environmental impact estimates were made using the Environmental Defense Paper Calculator. For more information visit: www.edf.org/papercalculator

Our Commitment to Green Publishing

Chelsea Green sees publishing as a tool for cultural change and ecological stewardship. We strive to align our book manufacturing practices with our editorial mission and to reduce the impact of our business enterprise in the environment. We print our books and catalogs on chlorine-free recycled paper, using vegetable-based inks whenever possible. This book may cost slightly more because we use recycled paper, and we hope you'll agree that it's worth it. Chelsea Green is a member of the Green Press Initiative (www.greenpressinitiative.org), a nonprofit coalition of publishers, manufacturers, and authors working to protect the world's endangered forests and conserve natural resources. *The Mystery of Metamorphosis* was printed on Natures Natural, a 30-percent postconsumer recycled paper supplied by Thomson-Shore.

Library of Congress Cataloging-in-Publication Data
Ryan, Frank, 1944-
The Mystery of Metamorphosis : a scientific detective story / Frank Ryan.
p. cm.
Includes bibliographical references.
ISBN 978-1-60358-341-1
1. Metamorphosis. 2. Evolution (Biology) 3. Hybridization. 4. Larvae. 5. Williamson, D. I. (Donald Irving) I. Title.

QL981.R93 2011
571.8'76--dc22

2010052618

Chelsea Green Publishing Company
Post Office Box 428
White River Junction, VT 05001
(802) 295-6300
www.chelseagreen.com

For my mother, Mary

Metamorphosis: The action or process of changing in form or substance, especially by magic or witchcraft.

—*The Shorter Oxford English Dictionary*

Metamorphosis: A profound change in form from [one] stage to the next in the life history of an organism.

—*Random House Webster's College Dictionary*

CONTENTS

PART FOUR: *The Molecular Age*

ACKNOWLEDGMENTS

To a scientist dedicated to the search for objective truth, ideas are precious things, never more so than in the intellectual contribution any one individual has made to the ever-metamorphosing wealth of knowledge that is the cherished common legacy of a discipline. To research a book such as this, one inevitably invades these domains, sometimes sharing, at other times questioning, the inspirations behind a lifetime of effort. My debts are many, for this book is an essential part of a sequence of other books, with all that I owe to those who helped me. Thus it is with humility and the utmost gratitude that I express my thanks for the generosity of time and spirit with which so many busy colleagues have welcomed my intrusion into their lives and work. My special thanks, inevitably, go to Donald Williamson, of the Marine Laboratory, the Isle of Man, and to his wife and family for their hospitality; to Professor Jonathan Wigglesworth, for the interview and documents concerning his father, Vincent, and for the hospitality shown to me by him and his wife; to Professor Lynn Margulis, the University of Massachusetts Amherst, to whom I owe so much already, and whose original suggestion led to my discovering Don Williamson's extraordinary line of experiments; to Professor Chris Curtis, London School of Hygiene and Tropical Medicine; to Professor Simon Maddrell, Department of Zoology, University of Cambridge; to Professor James W. Truman and his wife and colleague, Professor Lynn M. Riddiford, both of the Department of Biology, University of Washington; to Professor Luis Villarreal, Department of Molecular Biology and Biochemistry, and Director, Center for Virus Research, UC Irvine; to Professor R. Thomas Zoeller, the Biology Department, Morrill Science Center, University of Massachusetts. I am also deeply indebted to the following, who were generous with their time and assistance: Michael L.

Arnold, Associate Professor, Department of Genetics, University of Georgia; Henry Bennett-Clark, Zoology Department, University of Oxford; Professor David Bradley, the Department of Infectious and Tropical Diseases, London School of Hygiene and Tropical Medicine; Professor Donald D. Brown, Department of Embryology, the Carnegie Institution of Washington; Dr. Alan J. Chalk, University of Delaware, for the interesting exchange of ideas; David Champlain, Assistant Professor, Department of Biological Sciences, the University of Maine; Jonathan D. Cowart, Department of Biology and Marine Biology, the University of North Carolina at Wilmington; R. Henry Disney, Department of Zoology, University of Cambridge; Brian Furner, Librarian and Director of Information Service, the London School of Hygiene and Tropical Medicine; Professor Michael G. Hadfield, Kewalo Marine Laboratory, University of Hawai'i at Mānoa; Michael Hart, Associate Professor, Department of Biological Sciences, Simon Fraser University, Canada; Sebastian P. Holmes, initially the Laboratory of Marine Biology, Isle of Man and subsequently Liverpool University; Michael Locke, Professor Emeritus, Department of Biology, University of Western Ontario; John M. McCoy, Vice President, Discovery Biology, Biogen Idec, Inc.; François Mallet, Ecole Normale Superieure de Lyon; Dr. Claus Nielsen, Associate Professor and Curator, Invertebrate Department, Zoological Museum, Copenhagen; Mark G. Nielsen, Assistant Professor of Biology, University of Dayton, Ohio; Professor Trevor Norton, Marine Biology, the University of Liverpool; Rachel O'Neill, Assistant Professor, University of Connecticut; Professor Sidney K. Pierce, Professor and Chair, Biology Department, The University of South Florida; Loren Rieseberg, Professor, Biology Department, Indiana University; the late Professor Sir Richard Southwood, Zoology Department, University of Oxford; Linda Van Speybroeck, FWO Postdoctoral researcher, Ghent University; Professor Richard Strathmann, Friday Harbor Laboratories, Washington; Billie J. Swalla, Associate Professor, Biology Department, the University of Washington; R. B. Toms, Department of General Entomology, Transvaal Museum, Pretoria, South Africa; J. Carel von Vaupel

Klein, of Leiden University; Kaori Wakabayashi, Department of Biology, Toyama University; Graham B. White, University of Florida; Cheryl Whitehorn, Medical Entomology Technician, the London School of Hygiene and Tropical Medicine; James B Whitfield, of the Department of Entomology, University of Illinois. I would also like to thank the archives department at the *Caian*, Cambridge, and the Harvard University Archives. I am, moreover, indebted, if not in person but in inspiration to the work and writings of Sir Vincent B. Wigglesworth, Charles Darwin, and Jean-Henri Fabre.

It gives me the greatest pleasure to acknowledge the contribution of Joni Praded at Chelsea Green, whose belief in this book was unshakable. And what a wonderful opportunity it was to work with Jonathan Cobb, whose insightful editorial comments and suggestions made a very real contribution to the final text. My agent, Jonathan Pegg, was a pillar of support, as were the project manager, Pati Stone, and copyeditor, Cannon Labrie, and all others at Chelsea Green. It was a real pleasure to work with such an efficient group. As always I acknowledge the love and support of my wife, Barbara, invariably my toughest critic and true believer.

FOREWORD

Becoming is superior to being.

—Paul Klee

Metamorphoses have long held a spell over the human imagination. For the ancients, as chronicled by the Roman writer Ovid in his *Metamorphoses*, the gods could assume any and many forms, like the comic-strip superhero Metamorpho, able to shape-shift his entire being into a mere part of the human body. Drowned Narcissus is turned into a yellow-and-white flower, the narcissus. Phaethon's sisters are metamorphosed into poplar trees, their tears hardened into drops of amber. Jupiter transforms the nymph Io into a heifer to conceal her. Daphne eludes Apollo by changing into a laurel tree, the lovelorn god winding up with laurel leaves in his hair. Neptune turns Cygnus into a swan—and so it goes, one organism transforming into another, pushed by desire rather than regard for academically immutable species lines. In fact, one might argue that Aristotle, often called the first biologist, rejected natural selection because he associated it with the unsavory superstitions of his imaginative Greek forebears, especially Empedocles, whose early version of natural selection featured "man-faced ox-progeny" that must have reminded Aristotle more of myth than science. The fifth-century B.C. philosopher Empedocles of Acragas (now Sicily) devised a theory in which organs, arising outside bodies, first roamed the land, the less fit ones dying out and the remaining ones coalescing to make bodies. Charles Darwin knew about Empedocles' theory and its rejection by Aristotle, whose writings Darwin's friend translated so that he could print it in *The Origin of Species* to show the cluelessness of the ancients.

In keeping with Aristotle's throwing out the baby of natural selection with the bathwater of mythical hybrid beings, metamorphosis—

from the Greek words *meta* (μετα), "after," and *morphe* (*μορφή*), meaning "form"—continues, for the most part, to be considered fantasy in popular culture. Except for some details of insect biology, it is considered a kind of magic or reincarnation, a subject for fiction not science. In Kafka's *The Metamorphosis*, perhaps the most famous short works of fiction ever written, Czech clerk Gregor Samsa wakes up transformed into a giant cockroach; he is unable to roll over but still conscious of his human family, to whom he listens in the next room. Britain's Prince Philip is alleged to have stated that he wished to reincarnate as a deadly virus in order to reduce the world's population; and, in a television-age throwback to the strange comeuppances doled out by the gods, the womanizing character Quagmire, in the cartoon *Family Guy*, is reincarnated as a prophylactic.

However, metamorphoses are real. The striking metamorphoses of organisms are not confined to butterflies and children's storybooks, but apply to sea creatures, many species of insects, and arguably, in a muted or "incomplete" form, to we ourselves in the difficult, hormonally mediated transition from preteen to adult. Like Aristotle, Frank Ryan, an evolutionary biologist as well as a medical doctor, dispenses with the gods—but not with natural selection—as he explicates the "beautiful mystery" of nature's transformations. And he resuscitates the seemingly primitive idea that Aristotle dismissed, of radical hybrids.

Ryan discusses the personalities, by turns bold and retiring, social and introverted, confrontational and easygoing, of key figures in our emerging understanding of the chemical—and evolutionary—bases for metamorphoses. Ryan globe-trots to meet the key figures in this biological subfield who've piqued his—and through him, our—curiosity. When they are not alive, he talks to their students and family members; and an expert himself, a doctor who has pioneered new evolutionary concepts about viruses and written a bestseller on tuberculosis, Ryan reads the primary scientific literature before relaying the amazing findings and open questions to us. Here we encounter the greats of classical metamorphoses studies, from the observations of Jean-Henri Fabre, the nineteenth-century author of *The Life of the Caterpillar* to the appropriately named Vincent B. Wigglesworth, who pioneered our understanding of hormonal transformation in insects, the chemical essence of insect pubescence. At first glance a little-known story such as this one might seem of interest only to a small coterie of specialist academics. But Ryan shows that the extent of the research of Wigglesworth and his students reaches far and wide—into the manufactured paper of the *New York Times*, *Wall Street Journal*, *Boston Globe*, *Science*, and *Scientific American*, for example, which was shown to contain plant-produced compounds that stop fire bug larvae from undergoing metamorphosis.

The case of the book's other major protagonist, Donald Williamson, is wilder. For Williamson has offered a new theory of metamorphosis that to mainstream biology is as easy to dismiss as Aristotle was for Darwin—or Empedocles for Aristotle. Williamson iconoclastically proposes metamorphosis comes not of the "gradual accumulation of mutations" over millions of years, as neo-Darwinism would have it, but issues rather from "forbidden fertilizations"—the joining of genomes of completely different species in new, sometimes viably reproducing bodies.

Williamson's radical proposed explanation sounds less fantastic if you realize that many nonmammalian animals fertilize eggs outside their bodies. For example, it is completely within the realm of possibility that flying insects in the late Paleozoic swampy tropics dropped some of their sperm on the egg masses of velvet worms. Velvet worms, called *Onychophora*, abound in tropical leaf litter on the forest floor. They extrude gummy, sticky stuff to help them catch edible prey. Velvet worms are tropical, crawling, many-legged beings who would look like caterpillars to your mother. Most of the time, such inappropriate sperm drops by an oversexed impatient flutterer would have had no effects, but not always. Sometimes sperm and egg unions of very distantly related animals are neither infertile nor aborted. Williamson was impressed by the fact that juvenile forms of very different species can be amazingly similar given the extreme differences of the adults they become. Following his curiosity, he tried to interbreed different species of marine invertebrates: he fertilized eggs of an ascidian (*Ascidia mentula*) with sperm of a sea urchin (*Echinus esculentus*). In one experiment all the eggs developed as easel-shaped pluteus larvae—sea urchin larvae from ascidian eggs. About 8 percent of these plutei grew into sea urchins, but the other 92 percent resorbed their arms and became spheroids, each with a sucker. Why does this matter? Because no adult or larval sea urchin ever grows any head suckers—but the tadpole larvae of ascidians do. (Later experiments crossing the sea urchin *Psammechinus miliaris* with the heart urchins *Echinocardium cordatum* were recorded in the documentary *Hopeful Monsters* by Robert Sternberg of Imperial College, University of London.) In Williamson's view, the overlapping genomes express themselves with a time lag in development. The pluteus, tadpole, caterpillar, or other animal genes of the larval form are expressed first; then, exquisitely sensitive to environmental change, they shut off, and the adult gene expression begins. This *is* metamorphosis.

If accepted, Williamson's conclusions will require reorganization of the entire science of zoology. The most telling proof that Williamson is correct lies in the appearance of head suckers ("adhesive papillae") in two sets of his hybrids. Head suckers typically appear (in ascidians) during metamorphosis when fishlike swimming larvae transform into sessile bottom-dwelling adults. But now why would fish-tailed larvae, which resemble our vertebrate ancestors, transform into decidedly un-fishlike adults that produce eggs and sperm but, attached to rocks, never travel again? The situation is one of those tellingly imperfect would-be designs, like the wings of flightless birds, that clue us in to the fact that the organism before us is the result of peculiar and unexpected history.

Williamson's work suggests that the adult forms of metamorphosing species—the radially symmetrical sea urchin on the sea bottom, the frog leaping about the bank of a forest pond, the flighty butterfly fluttering and hovering in the air—are literally alien genomes with respect to the larval forms. Interspecies beings, these organisms go through such profound changes because they have successfully merged stranger genomes. Such Prince Charming transformations seem to testify to quite complicated ancient relationships; they are emblems of the most boggling sort of "mixed marriages" that, despite their starkly different backgrounds, somehow genetically worked it out.

Whether or not Williamson's ideas (wholly or in part) ultimately become accepted, the fact remains that, unlike the fairy tale in which the princess kisses the frog, these larval forms really do metamorphose. Readers of Ryan's book will find that these metamorphoses remain a mystery—but less so than when they started the book. Ryan makes the beauty of metamorphosis palpable, and lays down the work of those variant personalities—by turns taciturn and gregarious, methodical and intuitive—who've attempted to wrest from nature the secrets of ontogenetic transformation. Apart from the practical, philosophical, and evolutionary ideas broached and explored, Ryan's work is a model of informed scientific history. He has the advantage of being well versed in the difficult terrain that he traverses. But he is also able to write clearly, to synthesize broadly, while still resisting the temptation to foreclose or come to too-precipitous conclusions. In an intricate but always fascinating and detailed account of metamorphoses and the scientists who study them, Ryan presents both an ethnography and a history, not only of science as it is believed and accepted to have happened after the fact, but also of science in medias res—science right now, before our eyes, forming—emerging and groping toward new understandings as if it were itself some sort of metamorphosing being, rustling in its pupa before emerging resplendently into the adulthood and newfound symmetry of a deeper understanding of biological process and the natural histories that mark it.

— Dorion Sagan and Lynn Margulis

PROLOGUE

The Beautiful Mystery

It is a hot day at the end of August, and students at Amanda Elementary School in Manhattan, Kansas, stream out of classes to gaze heavenward. After several days of clouds and rain the weather has improved, and today the sun is beaming down through a clear blue sky. Suddenly, fingers point to where splashes of golden orange are soaring above the playground. The great monarch butterfly migration has begun. Farther south, at Emporia Village Elementary School, students are also staring up at the sky and counting. By the end of the day they will have already listed 100 sightings. Later that afternoon, monarchs come pay a visit at Corinth Elementary School in Shawnee Mission. Hundreds float high in the air, then all of a sudden descend on the playground, accompanied by birds and dragonflies that appear to be making the journey with them. The monarch migration is an exciting annual attraction for children in America, when up to 100 million butterflies migrate from the colder North to winter on the warm Californian coast or the Sierra Madre in Mexico. It is a journey as old as, and maybe a great deal older than, the great migrations of buffalo that have entered legend. On the Web sites that link to the school biology classes, teachers explain what is happening: the story of caterpillar and butterfly, two utterly different life-forms following their different life cycles. The children are fascinated to learn of how the monarch "changes its ecological niche entirely when it transforms from a caterpillar to a butterfly . . . a miraculous biological process of transformation [involving] two ecologically different organisms, as distinct as a field mouse and a hummingbird." It appears almost magical, a transformation that has captured the human imagination from classical times—and perhaps even longer.

We can admire the naturalistic animal paintings on the walls of rock shelters in Australia and South Africa, and deep within the gloom of French and Spanish caves, dating back to the Aurignacian period, roughly 30,000 years. Exquisite sculptures date from the same period, such as the figurines of a diving cormorant, or the putative head of a horse, carved out of mammoth ivory, that were discovered by the German palaeoanthropologist Nicholas Conard at the Hohle Cave in the Ache Valley.[1] One such ivory figurine, just a few centimeters long, depicts the hybrid features of a man and a lion. We can't be sure what the artist intended, but given that a second lion-man sculpture has been discovered in a nearby German cave, Conard suggests that this might fit with a common shamanistic religious experience, suggesting that "the transformation between man and animal, and particularly between man and felines, was part of the Aurignacian system of beliefs."[2] For Conard, this extension of creativity, from mere observation of nature, to a more ideational, perhaps spiritual inspiration, was the most exciting discovery of all. "I've been digging for a long time and I'm usually very calm in my work—but this certainly got my heart pumping."

There is a word for the dramatic transformation of one being into another. We describe it as metamorphosis. This is the title Ovid chose for his classic collection of exotic and wonderful stories, so reminiscent of what was seen in the ancient sculptures and cave paintings, of how men and women lost their human forms and souls to be transformed, by divine miracle, into four-legged animals, birds, and flowering plants. It seems not unreasonable to assume that such inspiration came from the everyday observation of the natural world, where the mystery of metamorphosis is all around us. Indeed, the more one looks to nature, the more widespread appears to be this phenomenon of bizarre, spectacular change. It extends far beyond the world of insects to include amphibians, such as the familiar frogs and toads, and in its most dazzling variety, to the strange and colorful marine invertebrate creatures that inhabit the hidden depths of the oceans, such as starfish, sea urchins, crabs, and sea squirts.

All such metamorphoses involve a series of transformations, so that the developing being exists in a variety of different stages, or forms, from the egg, through one or more "larval" stages, such as we see in the grubs or caterpillars of insects, or the bewildering variety of marine invertebrate forms, to a transformation that results in the final, or "adult" form of the insect or periwinkle, lobster, or salamander. An intriguing example is the startling metamorphosis of the starfish, *Luidia sarsi*, in which larva and adult coexist simultaneously as independent life-forms in the different ecologies of the surface waters and the ocean floor. If we did not know that the larva and adult starfish were born from a single fertilized egg, we would view them as radically different animals that belonged to completely different branches of the tree of life. In fact, this is no rare or fantastic freak: it's an altogether typical example drawn from the bizarre, entrancing, some might say magical, process of transformation the starfish shares with the majority of the animal species that inhabit the oceans, lands, and air of our planet. It is part of the mystery of metamorphosis.

In nature, we are familiar with the metamorphoses of moths and butterflies, where the humble caterpillar stops its frantic feeding to enter a phase of quiescence, known as the pupa, or cocoon. Walled off from the world, it appears to undergo a mysterious, and wholesale, transformation from which, as if by some quasi miracle, the glory of the adult eventually emerges. For the Dutch anatomist Jan Swammerdam, judged by many to be the foremost naturalist of the seventeenth century, the metamorphosis of caterpillar to butterfly symbolized the journey from pedestrian life to death and resurrection in the afterlife, with the pupa representing the repose of the soul between death and the Day of Judgment. Such ecstatic vision is perhaps understandable when we see what emerges from that seemingly deathlike cocoon, the newly emerging adult insect with its multifaceted eyes, articulated legs, newfound sexual maturity, and extraordinary wings. That two such very different beings could derive from a single fertilized egg is at once shocking and thrilling. It is little wonder that it has long intrigued the human imagination.

For a great thinker such as Aristotle, the butterfly's caterpillar stage was a continuation of its embryonic life that extended to the formation of the perfect adult. "The larva, while it is in growth," he argued, "is nothing more than a soft egg."[3] William Harvey, famous for his discovery of the circulation of the blood, saw some mystical influence at work in the transformation of the larva into an entirely new form. In the pages ahead we shall meet other naturalists who were equally enthralled, including Charles Darwin, the father of modern evolutionary theory, and the French "poet" of nature, Jean-Henri Fabre. A modern example is Don Williamson, a marine biologist who has spent his working life at the former Marine Laboratory on the Isle of Man. If Williamson is right, metamorphosis is even more intriguing than some of these past luminaries of science and philosophy dared to imagine.

Williamson's ideas have provoked such controversy in the fields of marine biology and entomology that some colleagues will object to my including his theory at all. I hope that, by the conclusion of this book, my readers will understand why I include him, in part because his story encapsulates this very conflict with orthodoxy, but more importantly because there are aspects of his theory that, taken on balance, offer enlightenment as well as the potential for new research. The mystery of metamorphosis is truly complex and profound—so much so that no single theory, whether orthodox or radical, can be considered in isolation. In addition to important contributions from Darwin and Fabre, I shall focus on the pioneering work of Sir Vincent Wigglesworth, widely regarded as the father of insect physiology, the American entomologist Carroll Williams, and the modern inheritors of both Wigglesworth and Williams's pioneering enlightenment, the husband-and-wife team of Lynn Riddiford and James Truman.

I don't pretend that science has all of the answers—metamorphosis still jealously guards many of its mysteries. But what we have captured to date is so precious and fascinating, it is no exaggeration to say that, even in its scientific exploration, metamorphosis remains both awesome and beautiful.

We shall examine these various scientific explorations of the mystery, which includes transformations in real life that would appear every bit as exotic and strange as those we encounter in Ovid's fantasies. The mystery of metamorphosis will be seen to lie at the very heart of the origins of the animal kingdom, extending beyond insects and the diverse and fascinating marine divisions of life to the vertebrate animals, including the familiar frogs and toads, and possibly to certain aspects of humanity. There are scientists who believe that we humans undergo metamorphosis in the profound and body-changing experience we are familiar with as puberty—and there are extrapolations of metamorphosis that may help us to understand the development of the very organ that defines our sentience, our extraordinary human brain.

PART ONE

Anomalies in the Tree of Life

Those strange and mystical transmigrations that I have observed in silk-worms turned my philosophy into divinity. . . . Ruder heads stand amazed at these prodigious pieces of nature, whales, elephants, dromedaries and camels . . . but in these narrow engines there is more curious mathematics. Who . . . wonders not at the operation of two souls in those little bodies?

—THOMAS BROWNE, *Religio Medici*

1

The Birth of an Idea

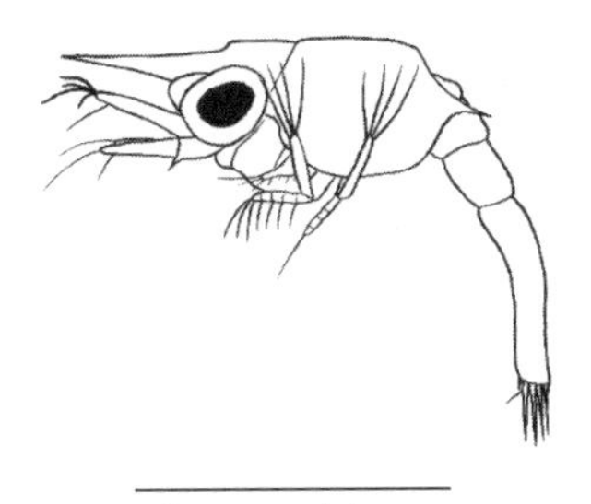

Sponge crab larva

The Northumberland coast, in the northeast of England, is as lovely as it is sparsely populated. Here, in the early 1930s, the youthful Donald Williamson would explore the lively fishing town of Seahouses, population little more than a thousand, where his father was a schoolteacher and amateur naturalist. At a time when the oceans had not yet been overfished and polluted as we find today, the herring boats would chug out to sea from the old harbor and return so overladen with catches that MacKay's fish shop would sell fresh herrings for a penny. To the north and south lay the great windswept beaches of Bambrugh, Beadnell, and Embleton Bay, where the lobster creels would be piled high on the harbor. On his walks, he came across bladder wracks to pop, or he collected the whorled shells of sea snails, some lined with brilliant mother-of-pearl inlays, and sea urchins and starfish with their exotic prickly shapes and colors. But it was the seabirds that most fascinated the boy. His greatest treat, once a year, was a trip with his father to visit the bird sanctuaries in the neighboring Farne Islands. Here he gazed spellbound at the masses of squawking gulls, ducked his head to avoid attacks by the dive-bombing terns, or was deafened by the hundreds of kittiwakes, all announcing their names with their songs in counterpoint.

In these surroundings Don Williamson fell in love with nature, and with the marine aspects of nature in particular, so much so that, in 1940, at the age of eighteen he enrolled for a bachelor's degree in zoology at King's College, Newcastle on Tyne, then part of Durham University. The Second World War had begun, and his first two years there included some stints of military training in addition to conventional courses in zoology, botany, and chemistry. In June 1942, he enlisted in the Royal Navy as a probationary sublieutenant, leading to training as a radar officer before serving in the Mediterranean theater on *HMS Abdiel* and *HMS Antwerp*. It was an eventful experience that ended, in March 1944, when he suffered a pulmonary hemorrhage en route to Taranto, Italy. He was taken to a military hospital where he was invalided out with the worrisome diagnosis of pulmonary tuberculosis, a disease that still awaited modern treatment with chemotherapy. After brief spells in several military and naval hospitals, Williamson ended up, far from his beloved sea, in Wooley Sanatorium, near Hexham in Northumberland. Here, in the sanatorium library, he came across a copy of Darwin's *Origin of Species*. In his words, "It was the first time I ever had the opportunity of reading it."[1] Inspired now more than ever to return to his biological studies, Williamson defied medical advice and returned to Durham University, where he completed his bachelor's degree while breathing on one lung, his other lung kept out of action through a surgically induced deflation, known as an artificial pneumothorax. Supported by a disability allowance from the Royal Navy, he went on to complete his PhD. The subject of his research was inevitably marine: a tiny crustacean known as a sand hopper, which, as its name suggests, hops about beaches in between tides.

Sand hoppers can easily be recognized by their small size and their shrimp-shaped bodies, bent into a C-shape, with their long antennae continuing the curve. In Australia these include the "bush fleas" and "leafhoppers" that are common in gardens and undeveloped bush—"small brown critters" that leap up when disturbed. In Britain, the land-based varieties are confined to beaches, with the exception of a single species that has invaded household greenhouses.

One species of sand hopper, known as *Talitrus saltator*, particularly intrigued Williamson and would become the subject of his doctoral research. He observed how it spent the daylight hours above the high-water mark, avoiding desiccation by hiding under the sand or decaying debris and only coming out at night to feed on washed-up wrack. He set out to explore the sensory cues behind their interesting behavior. At this time biologists believed that sand hoppers found their way by a kind of position sense that enabled them to detect the slope of the beach. Choosing a time when the tide was out, Williamson ferried his hoppers from the high-water mark to the low-water mark, then monitored them as they made their way back up the shore. How speedily and accurately they managed this, traveling upslope in what appeared to be straight lines! If they navigated using a sense of gradient, then vision was not the primary sense involved. But when he covered up their black multifaceted eyes, he discovered that they completely lost their way. He tested this further by showing them lantern slides of simple shapes and shades and confirmed that they were attracted by boundaries where dark met light. It was clear that the prevailing views were wrong. Sand hoppers navigated the beach through visual cues and not through sensing the gravitational pull of its slope.

Two years later, his dissertation complete, Williamson needed to find a job. He was offered a post in Jamaica, but it would oblige him to become an entomologist. The University of Sheffield offered an alternative, but the city was landlocked, and he would have to sacrifice any possibilities of working with the sea. He opted for the Isle of Man, where a post was open in marine zoology, and specifically that of planktologist with lectureship duties to Liverpool University. "Of course, I was not a planktologist, having conducted my research on hoppers, which were semiterrestrial. But nevertheless I applied for the job and I got it. I now had to learn about plankton."

The word *plankton* derives from the ancient Greek word for a wanderer. And how wonderfully apt it is! Imagine a planet where 99 percent of the living space is ocean. Of course we're living on it. Although we rightly assume that only two-thirds of the earth is covered by water, any true comparison must take into account biospheric volume: the terrestrial habitat is largely two-dimensional, while the oceans are three-dimensional and, in places, miles deep. The surface waters are inhabited by tiny open-water algae, known as phytoplankton, and minuscule animals, known as zooplankton, as well as a multitude of different bacterial forms that conduct their lives in a veritable zoo of unknown viruses. Plankton, in this minuscule plant, animal, and bacterial form, is the foundation of the oceanic food web, providing food for larger creatures such as fish and whales. Plankton also makes an important contribution to life in general. Phytoplankton, such as algae, live close to the surface, where, like plants on land, they capture the energy of sunlight, taking carbon dioxide out of the atmosphere and releasing oxygen back into it. In this way they play a key role in Earth's oxygen and carbon cycles, while also reducing the tendency to global warming. Others are involved in the nitrogen cycle, capturing nitrogen from the atmosphere, incorporating it into more complex organic chemical compounds, which are then fed back into the web of living interactions that form the essential basis of life on Earth, providing nutrients for all plants and animals. Zooplankton do not necessarily need light and so can live at any depth in the so-called pelagic, or upper reaches, of the oceans. They include an extraordinary diversity of forms, many of which display a spectacular if eerie beauty, best seen at low magnification and when using a transilluminating source of light. A major component of zooplankton is the larvae of marine invertebrates, including those same sea urchins and starfish Don Williamson had collected on the Northumberland beaches.

Marine larvae are an integral stage in the metamorphoses that encompass a wide variety of marine invertebrate animals. In his role as planktologist, Williamson would conduct original research studies on these larvae. But one group of marine creatures in particular would come to fascinate him: the crustaceans. The crustaceans constitute a major division of the animal kingdom known as a phylum, which, together with two other big groups, the mandibulates, which have antennae and jaws and include the insects, and the chelicerates, which lack antennae and jaws, make up the superphylum of the arthropods—invertebrate animals with jointed legs.

All crustaceans have a hard outer shell and two pairs of antennae adorning their heads. They include crabs, lobsters, and shrimps as well as a bewildering variety of less familiar creatures, varying from barnacles to water fleas. When, in the 1950s, Williamson first began to study crustacean plankton, he discovered that many areas remained to be explored. For example, the larvae of many of the hermit crabs found in British waters had never been described. Teaming up with a colleague, Richard Pike at the Millport Marine Biological Station on the Isle of Cumbrae, Scotland, he set about filling in the gaps. In time, Williamson became a globally acknowledged expert on marine metamorphosis and, in particular, on the larvae of crustaceans. But again and again, in describing such metamorphoses, he found himself confronted by mysteries.

Take, for example, the life histories of sponge crabs and hermit crabs.

Sponge crabs are so-called because they carry living sponges above their carapaces, as we might protect ourselves with umbrellas against the rain. The sponges provide the crabs with protection of a different kind, as a mixture of camouflage and protective shield in times of danger. If a predator threatens, the crab remains motionless under its living umbrella. If a determined predator still takes a bite, all it gets is an unappetizing mouthful of sponge, with prickly spicules. Sponge crabs are true crabs, slotting appropriately into the crab section of the tree of life—in scientific jargon, the "evolutionary" or "phylogenetic" tree. But the larvae of sponge crabs don't look like the larvae of the other true crabs. Instead they closely resemble the larvae of hermit crabs, which are not true crabs at all but are more closely related to lobsters. It struck Williamson as peculiar that, in spite of the considerable evolutionary separation between sponge crabs and hermit crabs, their larval stages were virtually the same.

He racked his brains to find a conventional explanation, but could not find one. "From then on," as he would subsequently recollect, "I wanted to solve the sponge crab paradox."

Conventional explanations of the unexpected larval similarities called upon a principle known as "convergent evolution." Take the familiar dolphins and whales. In their streamlined shapes and the use of their tails for locomotion, these marine mammals resemble fish. This is an example of convergent evolution. Moreover, it is easily explained. These marine mammals are rapid swimmers in the same environment as fish, the oceans, and so natural selection, which ultimately dictates the shape and movement of fish, has also adapted the shape and movement of marine mammals along similar lines. However, when one contrasts and compares marine mammals and fish, the convergence is seen to be superficial. In their internal anatomies and organs, dolphins and whales bear little resemblance to fish. Like all mammals, they are warm-blooded, and they use lungs to breathe oxygen from air rather than extract it from water, using gills. In many other aspects of their internal anatomy the superficial convergence does not mask their separate evolutionary histories.

Another striking example of evolutionary convergence is the similarity between the eyes of mollusks, such as squids and octopuses, and those of vertebrates, such as fish and humans. Both are camera-type eyes in which an image is captured by a lens and focused onto a light-sensitive retinal layer. The eyes are also remarkably similar in many other facets of their appearance and organization. The commonalities extend to a developmental gene, known as *Pax-6*, which plays a key role in constructing the eyes of squids and fish during their embryology. *Pax-6* appears to play a similar role in eye development throughout the entire animal kingdom, from the simple eyespots of earthworms to the multifaceted eyes of butterflies, and, remarkably, to the eyes that enabled the genius of Rembrandt. This would suggest that all visually endowed animals share a distant, likely very basic, common evolutionary ancestor that first discovered a means of responding to light. But does it mean that Rembrandt also shares the detailed evolution of his vision with octopuses and squid? A previous generation of biologists refused to believe so. They saw the present-day similarity in the eyes of mollusks and humans as a classic example of convergent evolution. And time has proven them right, though the proof involves degrees of subtlety that were only revealed with modern tools of molecular biology coupled with precise study of development, which led to the discovery that specific components of the eyes of vertebrates and squid develop through quite different mechanisms and from different embryonic sources. While we may well share a very distant common ancestor with mollusks, which might explain our common use of the *Pax-6* gene, the more overt similarities of our camera-type eyes have actually come about through much later convergent evolution.[2]

Convergence is thus an important, sometimes compelling, explanation of some parallel forms seen in nature. Williamson, then, posed the question: was convergence a convincing explanation of the striking similarities between the larvae of the sponge crab and the hermit crab?

Anyone who has studied marine larval forms cannot fail to be impressed by their amazing diversity of shapes and patterns. Fast swimmers, such as fish and whales, may have evolved a streamlined shape as a result of convergence through the need for locomotion in the same ecologies, but there is no discernible convergence of shape among the myriad larvae that inhabit the same pelagic streams of the oceans. This suggested to Willamson that similarity of ecological needs and constraints was unlikely to explain the similarities he was observing in larvae from two widely divergent places on the evolutionary tree. Indeed, the harder he probed the relationship between marine larvae and their place on the conventional tree of life, the more anomalies he appeared to discover.

In the late nineteenth century, scientists believed that the embryonic development of any creature recaptured its evolutionary history, a concept that was first proposed by the German naturalist Ernst Haeckel as a biological law. Today developmental biologists no longer believe this recapitulationist theory in an absolute sense—

evolutionary adaptation can occur at any stage, including the embryo and larva—but still it can be useful, if treated cautiously. For example, the fertilized human egg develops into an embryo equipped with a tail and fishlike gills. Evolutionary biologists believe that a distant ancestor of mammals, including humans, was a fishlike animal that possessed gills and a tail. The human embryo could be portrayed as recapturing this stage of our evolutionary past. Many evolutionists invoke Haeckel's law to some degree in attempting to explain the changes of metamorphosis. However, Williamson saw a flaw in such evolutionary thinking when it came to the important marine phylum of prickly-skinned animals, such as sea urchins and starfish, known as the echinoderms.

Part of our fascination for these exotic creatures lies in their rounded and starry shapes, which, like flowers, imbue them with an aesthetic beauty that is so unlike most of the animals we see on land. We humans have a left and a right side. We are bilaterally symmetrical. Echinoderms don't have a left or right side; rather, like the petaled heads of the majority of flowers, they are radially symmetrical. How strange, then, that radially symmetrical sea urchins and starfish begin their lives as larvae that are bilaterally symmetrical. This change during development, from bilateral to radial symmetry, involves one of the most spectacular metamorphoses in all of biology, and it can only be brought about through wholesale reorganization of the larval anatomy, including skin, skeletal structures, the vascular circulation, and the structure of the nervous system. We shall look at the metamorphoses of starfish and sea urchins in more detail later, but for the moment it is only necessary to grasp the general principles.

Williamson found it hard to imagine how a radial starfish could possibly have evolved from a bilaterian ancestor. To his mind it seemed to imply that at some stage in its evolution the ancestor of the starfish had wiped out its inherited body plan and evolved a completely different development from scratch.

In the early days of his appointment to the marine station, he had not been required to lecture to students. But from the 1960s onward an increasing number of biology undergraduates began to take an interest in marine biology. Undergraduates who wanted to major in marine biology spent a year on the Isle of Man studying at the marine station. As part of a course he developed for these students, Williamson gave a lecture on marine metamorphosis and its larval aspects—essentially the same lecture, year after year. Then, as now, he was a firm believer in Darwinian evolution. He would tell his students that, in general, most marine larvae appeared to fit with what would be expected from orthodox evolutionary theory, but there were anomalies that demanded explanation. Most taxonomic classifications were based on the adult forms, with the larvae assumed to complement the positions of their respective adults on the tree. The problem, as Williamson had now come to realize, was that many larval forms just did not fit in with the extrapolation of the tree of life based on the adults.

Darwin had puzzled over the mysteries of metamorphosis and, as we shall see later, he had himself put right the misclassification of barnacles through a study of their larvae. Indeed, in *The Origin of Species* he concluded that an animal species should be seen as comprising all phases of development from the egg to the adult, each phase, including the larvae, undergoing its own separate evolutionary modifications under the influence of natural selection. The pressures of natural selection working on each of the different phases of the life cycle adapted the relevant form, whether larva or adult, to survive in its specific ecology. Or to put it another way, each phase of a complex animal life cycle, from embryo to larva and from larva to adult, would be honed by evolutionary pressures, with ultimate survival to the stage of successful reproduction—usually encompassed by the adult stage—the ultimate key to evolutionary success. In this way the free-swimming bilaterian larva of a starfish had been honed by natural selection to fit life in the surface, or "planktic," layers of the oceans—its key role being dispersal in as wide a geographic range as possible—meanwhile the radial adult had been honed to fit its predatory existence in the much more geographically restricted ocean floor. But while Williamson accepted that natural selection would separately adapt larval and adult developments, he saw many anomalies in this ideal simplicity—anomalies of symmetry, of detailed development of key organs and whole animal form—that were too readily dismissed by colleagues. He drew up a new tree of life for the marine invertebrate animals, based not on the adult body forms but on the various larval body forms, and he then compared his larval tree with the traditional tree based on the adults. If the larvae and adults had followed the same evolutionary trajectories—as Williamson assumed they must in orthodox theory, which implies a linear evolutionary development, including all stages of the life-form—the two trees should exactly overlap. But Williamson's larval-based tree did not overlap the orthodox adult-based tree.

"On occasions during evolutionary history," he would say to his students, "larval and embryonic forms that have originally evolved in one lineage have somehow appeared in another. It is as if they jumped from one branch of the phylogenetic tree to a distinct and sometimes distant one. Of course," he felt obliged to add, "such a thing could not have happened. The whole theory of Darwinian evolution stands firm against it. And therefore it is nonsense."

But it was a nonsense that intrigued him year after year.

In the autumn of 1983, when he was revising the same lecture for the twenty-eighth time, he dared to ask himself a shocking question: "What if it isn't nonsense at all? What if larvae really have moved across the evolutionary tree between species, genera, and even the massively different taxonomic groups known as phyla?"

Williamson tore up his old lecture notes and drafted some new ones. That year his students heard the first crude explanation of one of the most extraordinary evolutionary theories that have ever been proposed. "I delivered the lecture in November that year and I have never had such attention to a lecture before."

2

A Puzzle Wrapped in an Enigma

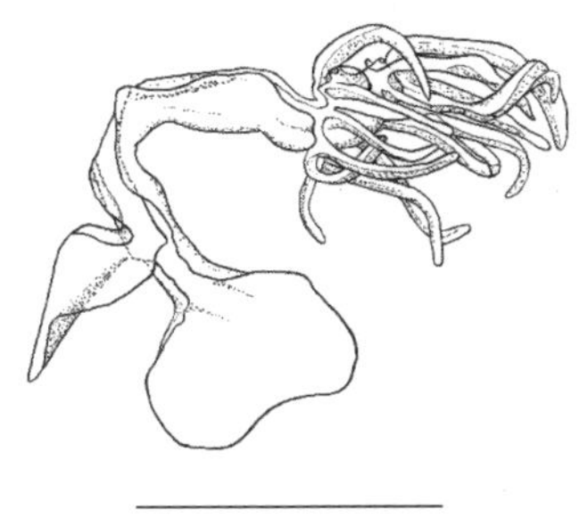

Larva of *Luidia sarsi*

Williamson decided that that he would take a long hard look at the world of marine zoology, looking for evidence that would back up his new idea of "larval transfer." And where better to look for anomalies than the familiar world of the echinoderms? These spiny-skinned creatures constitute a phylum all to themselves, including the starfish and sea urchins. Adults in all of the member species have three things in common: they are radially symmetrical, based on a five-segment or pentaradial plan; the various permutations of their internal skeletons and their stiffening spines, or external skeletons, are all made of calcium carbonate; and they have an internal body cavity, known as a coelom, that is divided up in a distinctive manner. Now, as Williamson examined their life and evolutionary histories in detail, he discovered a great many anomalies. For example, he looked in detail at the life history of the starfish, *Luidia sarsi*.

This animal, which appears to have no common name, is unusually large for a starfish, growing to eight inches in diameter. Ranging from sandy to salmon pink in color, it has the typical radial body plan of the echinoderms, with five gently tapering arms. Biologists describe its skin as having a velvety texture, and they remark on a conspicuous band of white spines running along the edges of the arms. *Luidia* inhabits the muddy sediment of seashores from Norway to the Mediterranean, hunting down its prey in the dark of night. But where most starfish feed on scallops and other bivalve mollusks, *Luidia* preys on other species of starfish, a fact that is pertinent to the more discerning scallop fishermen.

Luidia's larva is a diaphanous sprite that grows to an inch and a half long, the largest marine invertebrate larva Williamson has ever seen. The technical name for its body plan is "bipinnarian." In appearance it resembles an uprooted vegetable, with a tangle of roots at one end and two broad and fleshy leaves at the other. In fact the roots are ciliated arms, encircling the separate openings of mouth and anus, and the leaves function as the equivalent of locomotory wings. At night it feeds on algae in the surface waters, but during the day it sinks deeper, swimming through the pelagic layers with flapping undulations of its wings. Unlike the adult starfish, with its radial symmetry, the larva is bilaterally symmetrical. The adult is conceived—I cannot think of a more appropriate term—from a cluster of cells lining the internal cavity of the larva, and here it grows and matures, an alien existence independent of, and seemingly oblivious to, the larval body structures, axis, bilateral symmetry, and form, imbued with what can only be described as a complete disregard for every embodiment of its larval stage of existence. In time the juvenile starfish emerges from the still-swimming larva to settle on the ocean floor, where it begins its own independent existence hunting down other starfish for food. In the case of most other starfish the larva is sacrificed as a brutal conclusion to the metamorphosis. But in the case of *Luidia*, fate is kinder. After the adult has broken free of its tissues, the larva carries on with its independent lifestyle, swimming the pelagic waters and grazing on its vegetarian diet of algae.

This metamorphosis is so startling we are obliged to take a mental step backward and reflect on what we have witnessed. Two separate beings, the larval and adult starfish, with radically different body forms and ecological life cycles, have developed out of a single fertilized egg. When first he read about *Luidia*'s extraordinary metamorphosis, Williamson's curiosity was aroused.[1] In his words, "These facts needed an [evolutionary] explanation, but I thought at the time that it was up to the echinoderm specialists to provide one." No explanation had ever been profferred.

The dual development of *Luidia* takes place in a precise sequence: First we have fertilization and the original conception, with the larva developing from the fertilized egg. Then, within the larval abdominal cavity, a new conception appears, with the adult developing from totipotent cells—the equivalent in humans of the stem cells that are capable of giving rise to any body tissue. For a time two different life-forms coexist within the one body. What, in philosophical terms, is the role of the larva? It is not quite the same as a parent—the true parents of the adult are the same as the larva. Perhaps it would be more accurate to visualize it as surrogate parent. Moreover, the larval genome, present in every cell of its bilaterally symmetrical body, is identical to the genome in every cell of the radially symmetrical adult starfish. The differences seen in the two different body plans and life cycles do not derive from differences in the genes of the larval and adult cells but from the presence of two quite different developmental blueprints within the same genome. To put it another way, we need to explain how the separate evolutions of larva and adult, with their separate developmental programs, have entered the life history of a single starfish.

Conventional Darwinian theory assumes a linear process of adaptation, through natural selection, in the evolutionary history of an organism. It's not surprising that Williamson felt obliged to ask how such conventional theory could possibly explain the utter disregard of the larval development and body plan in the adult developmental blueprint. Would a linear process of adaptation not lead us to expect that the overall body axis and symmetry of the larva should be retained in the adult? Would we not expect the development of vital organs, such as the mouth and gut, however modified, to be inherited from larva to adult? Would we not expect the same process of adaptation and modification to apply to internal organs and tissues?

Williamson doubted that conventional evolutionary theory *could* explain the metamorphosis of *Luidia sarsi.* And if it failed to explain the more bizarre and wonderful, how could it possibly offer a comprehensive explanation of the full range of marine metamorphosis?

For all the variety of terrestrial animals, the splendor of life in the oceans is breathtakingly more diverse. And where metamorphosis, in remarkable form and beauty, is a feature of two terrestrial classes, the insects and amphibians, it features, in a bewildering compass of richness and variety, in at least fifteen separate marine phyla. The mystery of marine invertebrate metamorphosis is inevitably more challenging. At the same time, the solution to this great puzzle might itself hold the key to a very great prize. Life began in the oceans. And marine invertebrates were the first animals to evolve. Since marine metamorphosis is very ancient, perhaps even primal, then understanding marine invertebrate metamorphosis might open a new window of enlightenment onto the evolutionary origins of the animal kingdom as a whole. It is little wonder, therefore, that this mystery has engaged and baffled a great many minds. We might start by asking a fundamental question of our own: Why should marine invertebrate metamorphosis have evolved at all?

Five hundred million years ago, before the rise of the fish, ancient cephalopods dominated the oceans. The term *cephalopod* means a combined "head-and-foot," but even this bizarre appellation belies the marvels of form and life history this class of animals embraces. There are six hundred species of cephalopods, which belong to the most diverse of all marine phyla, the mollusks. Those ancient cephalopods are thus related to the humble snails we see in our gardens. With their jet-propelled locomotion, driven by three hearts pumping blue blood, their ability to change color faster than a chameleon, and their highly developed eyes—as large and sensitive as those of any mammal—this varied class includes the chambered nautilus, the flashback cuttlefish, the oceanic squid, and the fastest and most intelligent invertebrate animal on the planet, the octopus.

Among the cephalopods, the family of octopods takes its name from the fact that its member species possess eight arms, commonly misinterpreted as tentacles. The most familiar of these are the hundred or so members of the *Octopus* genus, which inhabit the ocean floor. In these animals we find that the typical mollusk shell is either lost or internally reduced. *Enteroctopus dofleini* is the largest octopus in the world, with individuals growing up to 150 pounds. This prodigious animal inhabits the continental shelf of the North Pacific Ocean, ranging from southern California to Alaska and extending across the Aleutians and into Asia as far south as Japan. The mature female lays her 20,000 to 100,000 eggs on the inner side of a rocky den, tending, cleaning, and aerating her brood until they hatch, between 150 to 210 days after fertilization. And here we witness a poignant example of maternal sacrifice. This devoted mother does not eat while she cares for her young, and she dies of malnutrition when the eggs have hatched. The hatchlings swim toward the surface, where they join the exotic zoo of planktic life-forms.

Though often loosely called larvae, octopus hatchlings are not larvae at all. In fact octopuses, like all cephalopods, do not undergo metamorphosis. Instead they hatch as miniature adults. These minuscule octopuses inhabit the planktic layers for 4 to 12 weeks, drifting with oceanic currents, until they reach a mantle length of about half an inch, at which time the young settle back to the bottom to begin their adult life history, searching out a new den and growing into fearsome predators of crabs and scallops.

All authorities agree that the cephalopod mollusks inhabit the two ecologies common to most marine invertebrates—an early spell exploring the free-floating three-dimensional ecology among the plankton of the surface waters followed by the mature life cycle in which the adult octopus must crawl and hunt in the two-dimensional ecology of the ocean floor. So how have they escaped the need for the dramatic changes of metamorphosis?

One of the most elegant ideas in embryology of the nineteenth century was undoubtedly Ernst Haeckel's recapitulationist theory. He also argued that the shapes and changes the embryo went through were of such critical importance to every animal that they must predict the development of the form and structures of the same animal throughout all of its subsequent evolution. This was widely believed in Darwin's lifetime, and it was extended to larval development. If the life cycles of marine invertebrates followed Haeckelian principles, there was no mystery about their metamorphosis. At some time in the distant past the ancestor of the species resembled the larva. This ancestor must somehow have evolved, in the linear manner proposed by Darwin, until it was transformed into the adult animal, with the remarkable life cycle we see today. In this way, every variety of marine metamophosis, including the most bizarre and wonderful, would be a palimpsest of its evolutionary origins and journey. While most modern biologists have rejected recapitulation as a universal law, sometimes the early stages of development really do give us useful information on ancestral form and structure. For example, whales don't have legs, but they have tiny remnant bones buried in their bodies, and leg extremities begin to develop in the whale embryo before receding again. Thus if we accept that recapitulationism is far from absolute, and if we can figure out the limited circumstances in which it actually applies, the study of metamorphosis might help biologists reconstruct some important aspects of evolutionary history.

In the 1870s, Francis Maitland Balfour, brother of the British prime minister James Balfour, took on this challenge while studying marine larvae at the University of Cambridge. He was one of the first to challenge Haeckel's theory, concluding that even the earliest embryonic stages of an animal's development were capable of evolutionary change, which would obscure or mislead any recapitulationist interpretations. He also concluded that evolution could suppress earlier developmental stages, so that the larval phase could be lost from a previously metamorphic animal life cycle. This implied that the interpretation of larval-adult relationships was trickier than ever. In an effort to explain the admittedly complex world of larval and adult relationships, Balfour made the bold assumption that metamorphosis was universally present among the earliest animals. To explain why octopuses and many other present-day animals do not undergo metamorphosis, he took the view that these must be descended from ancestors that actually did metamorphose, but had abandoned their larval stages over the eons of their evolutionary history.

Balfour decided that he would call the original and most ancient of larval forms the "primary" larvae. If these had evolved in linear Darwinian fashion from the earliest animal embryos, he believed the forms of these primary larvae would offer tantalizing evidence of the evolution of the earliest animal forms. The relevant metamorphosis would repeat its own ancestral history, and this would allow him to unravel some of the complex links and relationships that gave rise to the earliest evolution of the animal tree of life.

But when he searched for primary larvae he encountered another problem: even if he correctly identified a larva as primary, its original form might be greatly changed as a result of evolution affecting the larva itself. He realized that he would have to take his searches a major step further, comparing larval forms across a phylum, and even across the whole of the animal kingdom. Only then could he peel away the work of evolution to reveal the true ancestors of phyla, and, he dared to hope, perhaps even the primal ancestor of the entire animal kingdom.

Balfour and Haeckel were contemporaries, members of a global intellectual elite that, much as the artists, writers, and composers of their day, interacted with and listened to one another, leading at times to mutual enlightenment, and at other times to furious disagreement. In the early 1870s, Haeckel had proposed, on the basis of his own theory, that the primal ancestor of all animals was a very simple marine creature, resembling the earliest stage of embryonic development—a hollow ball of cells with a primitive mouth opening into a saclike gut. Haeckel called this the "gastraea."[2] Balfour decided that he would examine embryonic and larval forms throughout the animal kingdom, looking for the hard evidence of Haeckel's gastraea. He would also compare present-day larval forms with living and fossil adults in different groups. Where he found common form and structure in both larva and adult, he would assume that the present adult must be closely related to the ancestral stock of the group in which the larva was found. And any such larva could then be regarded as "primary."

But when he began the actual searches, the picture became increasingly complex and uncertain. Indeed, in time he arrived at the startling conclusion that most of the present-day marine larval forms, such as the larvae of the starfish and sea urchins, were not primary at all. They had somehow invaded the preexisting life cycles of species. Extrapolating from his theory, he then had to assume that in the distant past there would have been primary forms in these invaded life cycles but those primary forms had been lost and replaced by what he now termed "secondary" larvae. And while primary larvae might offer important clues to the distant ancestors of the present adult life-forms, these secondary larvae, as far as he could determine, provided no ancestral link at all to the adult phyla in which they were found.

The phylum of the cnidarians, which includes the sea anemones, jellyfish, hydras, and corals, became the main focus of his attention. These are the simplest of animals in terms of body blueprints and tissue organization, and so it will come as no surprise to find that these are also among the earliest animals to appear in the fossil record. They follow two basic body patterns: sometimes they are fingerlike, for example, the hydra, and sometimes they are medusoid, such as the jellyfish. Interestingly, both these body forms are radially symmetrical in the horizontal plane. Their larvae adopt the simplest of all larval forms, a radially symmetrical solid ball of cells, known as a "planula," which has a surface layer covered by locomotory cilia. Balfour saw the planula as a perfect example of his primary larva. If a medusalike radially symmetrical organism was one of the earliest animals on the scene, and if this animal had evolved from the still simpler planula, then the planula—that simple ball of cells, with its skin of beating cilia—was as close as he was likely to get to the primal animal ancestor.

What did his contemporaries make of Balfour's pioneering ideas, which he published in a book in 1881?[3]

Charles Darwin, a friend and admirer, wrote a letter to Fritz Müller, dated January 5, 1882, in which he stated:

> Your appreciation of Balfour's book has pleased me excessively, for though I could not properly judge of it, yet it seemed to me one of the most remarkable books which have been published for some considerable time. He is quite a young man, and if he keeps his health, will do splendid work. . . . He is very modest and very pleasant, and often visits here and we like him very much.[4]

However, on February 13, 1882, Darwin wrote a new letter to Dr. Dohrn:

> I have got one very bad piece of news to tell you, that F. Balfour is very ill at Cambridge with typhoid fever. . . . I hope that he is not in a very dangerous state; but the fever is severe. Good heavens, what a loss he would be to Science, and to his many loving friends![5]

This presaged a sad, and somewhat ironic, ending to this story. Balfour recovered from his typhoid fever and traveled to Switzerland for a period of convalescence, but, unfortunately, while attempting to climb the unconquered Aiguille Blanche slope on Mont Blanc, he died from a fall. He was just thirty-one years old. And as Darwin anticipated, the young naturalist's death, barely a year after publication of his pioneering ideas, was a major loss to the world of science, and to the burgeoning science of embryology in particular. How poignant also that on April 19, 1882, just two months after writing his letter of concern, Darwin himself died.

For Brian K. Hall, a professor in biology at Dalhousie University, Balfour's conclusions, however speculative, still provide a useful guide for continued embryonic study even today.[6] But by and large the broader world of biology soon forgot the pioneering work of Balfour. It was a neglect that would prove costly to Williamson, when, unaware for many years of Balfour's earlier work and thinking, he struggled to come to terms with similar questions and problems.

3

First Experiments

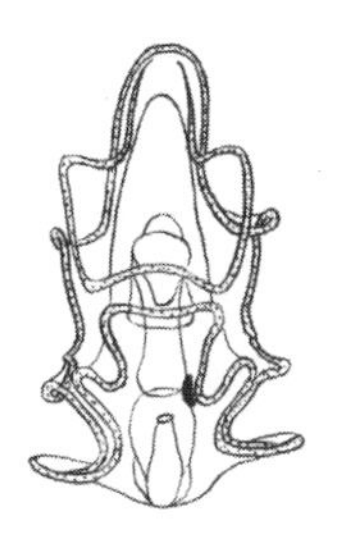

Bipinnaria larva of a starfish

The Port Erin Marine Laboratory on the Isle of Man was founded in the nineteenth century by a committee of amateur naturalists, headed by Professor William Herdman, a zoologist from the University of Liverpool. It began with two small buildings, but Herdman soon convinced the government that it was in their interest to expand the facility to include a fish hatchery, a place to study marine life, and a public aquarium. The government put up the money to build a more comprehensive institution, set amid the rocks of a quarry on the beautiful sweep of bay. At the time I first visited it, in 2002, the marine lab had long since been incorporated as a teaching group into the School of Biology in Liverpool. Students came over from Liverpool as part of the modular course in biology, and those with a deeper interest in marine biology spent their honors year working with the lecturers and postgraduates conducting various research programs. I was surprised to discover the global nature of this research, which was often highly practical—for example, helping colleagues to set up their own laboratory in Egypt or showing them how to manage fisheries in Chile, or a coral reef conservation program in the Philippines. Part of the remit was environmental awareness.

The Isle of Man has a central position in the Irish Sea, which has direct continuity with the Atlantic Ocean. The laboratory has conducted a continuous series of measurements, beginning in 1900 and extending to the present day, that have helped to confirm the steady increase in oceanic pollution and a slight, but significant, increase in oceanic temperature of 0.5 degrees over more than a century. Given the vast capacity of the Atlantic Ocean, this is important confirmation of global warming.

On a sparklingly fine day in January I found myself outside the main doors to the marine lab, listening to a little of this history and gazing across the bay of Port Erin in the company of Professor Trevor Norton, its current director. Gulls shrieked and nested in the cliffs immediately behind us. The headland on the far side of the bay was blooming with color even in the depths of winter. Around the base of a stone tower spread the golden yellows of European and Hispanic gorse, the gingery rust of bracken, the purple mist of heather. "We had two new students, one from China and one from the Philippines, and they asked me: 'Who planted all these flowers?'" Trevor threw back his head and roared with laughter. "As an Irishman you'll understand—there's magic in it."

Magic captured my enchantment exactly.

Professor Norton combines a love of science with the art of writing. He has written three delightful books about his memories of marine biology. These include *Reflections on a Summer Sea*, which is based on his youthful experiences among a menagerie of dedicated English ecologists working through a series of summers around Lough Ine in the southwest of Ireland. In reading the book, I was amused though hardly surprised—I grew up in England from my early teens—to discover that the English visitors to Lough Ine were themselves seen as eccentric by the native Irish. "In fact the scenery here," he informed me as we faced the decorous headland, "is very like that of Ireland. It is full of these lost beaches and beautiful forgotten valleys."

When he first formulated his hypothesis of larval transfer, Williamson realized that it was not enough merely to discover anomalies that might be explained by his hypothesis: he needed to find a plausible explanation for how a larval form might be transferred from one branch to another on the tree of life. Of course, he was not thinking that a larva had literally crawled or swum across the evolutionary tree. The transfer he had in mind was not of a whole organism but of the genetic blueprint for that larval shape and life cycle, which would need to be carried, through an evolutionary mechanism, from one life-form to another. He considered various options before concluding that it must occur through the biological phenomenon known as hybridization, which involves the sexual union of parents from different species. Hybridization, in bringing two different genomes into the genetic inheritance of a single offspring, offered the only plausible evolutionary mechanism for his theory of larval transfer.

One of the beautiful if unfrequented local beauty spots referred to by Professor Norton is Scarlett Point, a projection of gray carboniferous limestone ledges, broken through by vertical stacks of dark volcanic lava. A curved grassy path leads to a series of brackish pools out of reach of the tides on calm days but drenched in spray when the weather plays rough. These pools are inhabited by an unusual shrimp, *Gammarus duebeni*. This was the animal Williamson chose to study in his first hybridization experiment.

Gammarus means "lobsterlike." If this creature has a biological claim to fame, it is the fact that it is the only British species of shrimp to breed in a wide variety of salinities, from freshwater to highly salty, so that it is commonly found in pools subject to evaporation around the high-water splash zone. The male *Gammarus* is larger than the female and has specially adapted "toothed feet." He uses these to grip the carapace of the female when they get together, so she dangles in a love-hug below him. They remain paired in this intimate way for days. After mating, the female molts and lays eggs into her brood pouch, which cues the male to squirt his sperm into the same chamber. In many cases, the female will not lay her eggs unless she is being carried by a male. Later, when I questioned him, Williamson was a little vague in recounting why he chose this particular shrimp to conduct his first experiment. "I can't remember how I got the information—whether I did it by trial and error—although *Gammarus dubeni* will usually lay eggs when it is coupled with the male, it will also lay eggs even when isolated. Perhaps I was influenced by the fact it is extremely hardy, thus likely to survive the interference of prolonged biological experiment. And I knew where to get some."

In 1986, equipped with his daughters' shrimp nets, he made his way to the pools at Scarlett Point, where he fished for the shrimps. They weren't easy to see in the dark-reflecting water, camouflaged a nondescript grayish brown, and, even as full-grown adults, no more than two centimeters from antennae to the tips of their tails. Amphipod crustaceans, such as *Gammarus*, do not undergo metamorphosis, developing from egg to adult without undergoing a larval stage. This was the key to what he had in mind. Indeed it was an experiment that had never been performed before in the history of marine biology. He would see what happened when he mated this marine shrimp, which had no larva, with a sea urchin that metamorphosed through a larval stage. "So I set up a tank in my room at the marine biology laboratory, with *Gammarus dubeni.*"

He was pleased to find that the shrimps bred in the tank without any problem. Soon he had hundreds of *Gammarus* ready to begin his experiment. He began to isolate those females that seemed ready for egg laying. This he could detect because they would first of all molt. At this time the marine laboratory was equipped with a public aquarium that contained plenty of sea urchins. Williamson fished out two or three of the common British species, *Echinus esculentus*. Globular in shape and a bright pastel pink in color, this sea urchin grows to six inches on rocky shores, where it grazes on algae and encrusting animals. He didn't know how to distinguish male from female among the sea urchins, but he knew that if you inverted ripe specimens over a beaker they would reveal their gender by laying either sperm or eggs. This he now accomplished in large quantities without harming the creatures. Soon, he had what he wanted, a copious supply of sea urchin sperm. He placed female shrimps into a concentrated suspension of sea urchin sperm and waited for them to lay their eggs.

It took no more than a day or two for the female shrimps to start laying. But now he was faced with the problem of getting hold of the fertilized eggs, which were hidden away in the brood pouch under the thorax. Using two needles, he pried open these private quarters and teased out the eggs without damaging the shrimps. In conventional biological thinking, there should be no offspring from such a disparate sexual union. Hybridization usually involves crosses between closely related species. However in this instance, the shrimp, *Gammarus dubeni*, belongs to the phylum of the crustaceans while the sea urchin, *Echinus esculentus*, belongs to the phylum of the echinoderms.

Animals and plants are separated into distinct kingdoms in the taxonomic classification of life. Phyla are the next great divisions below those of the kingdoms. All the members of a phylum have some defining characteristic in common, often a characteristic that is very basic indeed. For example, we humans belong to the phylum Craniata, from the Greek *kranion*, which means that all the phylum members come equipped with a skull. The craniates, together with the phyla of urochordates and cephalochordates, make up the super phylum of chordates, all of which possess a nerve cord within a flexible spinal column. Within the craniates, we humans belong to the class of mammals, which contributes just 2,500 species to the 45,000 species of craniates, the latter including fish, amphibians, reptiles (including dinosaurs), and birds. The differences between phyla were laid down more than 500 million years ago, back in the evolutionary period known as the Cambrian, when the major branches of the animal tree evolved. In other words, members of different phyla are separated from each other by enormous distances in terms of evolutionary time, as well as by major differences in their physical and genetic makeup. A single observation will bring home the extent of the differences between the parental species that Williamson was bringing together in his hybrid experiment. *Gammarus dubeni*, like all Crustaceae, has a left and right side. It is bilaterian. Sea urchins, like all of the members of the phylum Echinodermata, are radially symmetrical.

Each morning, and several times during the day, Williamson would pore over the eggs with the aid of a microscope, looking for evidence of development. Many survived for no more than a day or two, but a minority survived anything from a few days to a week. An even smaller minority survived for more than a week. The outer membrane of shrimp eggs is a good deal less transparent than the eggs of echinoderms, and, however carefully he scrutinized the eggs, he could not make out cell divisions in the embryos. But after a time he could see that there was a spherical form in two of the eggs. He could also make out that the embryo was spinning rapidly, in a way that suggested it was being propelled with cilia. Crustaceans like *Gammarus dubeni* do not possess cilia, at embryonic or any other stage. This means that the normal shrimp embryo should not be mobile. So the movement of this tiny unknown offspring had to represent some unusual development, most likely a ciliated offspring.

He waited to see if either of the two eggs would hatch out, so he could inspect the resulting free-swimming life-form. But neither hatched. He attempted to dissect off the egg membrane without any success. When he tried to release the spinning form for dissection, whatever was inside disintegrated as soon as the egg membrane was ruptured, leaving him with an unrecognizable pulp. He considered the possibility of photographing the larvae inside the eggs. He had access to photomicrographic microscopes, but the movement within the eggs was so rapid, all he could see was a blur. Cine film photography down the microscope was not available to him at that time, so he could not record it. "It was frustrating but at the same time the results of this single experiment gave me great encouragement. I thought I really am on the right lines and hybridization is the answer. This must be the way in which some animals acquired what appear to be the wrong larvae."

People commonly assume that new scientific theories arise in the minds of their inventors much as we come across them in lectures, articles, and textbooks long after the theory was first conceived. In reality, nothing could be further from the truth. New scientific ideas, much as original creativity in painting or writing, are more likely to arise as inchoate ponderings, triggered by some flash of inspiration or some unlikely observation. This is exactly what we see with Williamson, for whom a new line of thinking was triggered by anomalous findings in a routine field of work. Scientists also tend to be cautious in their reactions. How many potential breakthroughs never come to fruition because their originators lose their nerve at the first hurdle? The temptation will be to put their new ideas aside for a while—witness how Darwin pondered his evolutionary theory for decades before he published it for a wider appreciation. In the early stages scientists are likely to search around to see if, maybe, somebody else has got there before them. If they encounter silence, they may gain confidence that their idea really is an original one and decide to take things further. They become sufficiently brave to go test their ideas, and sometimes this bears fruit. Through such predictably human mixtures of inspiration, fearful uncertainty, and thrill of exploration creative ideas eventually emerge into the light of day.

And now, however frustrating the shrimp-urchin experiment, Williamson had a better grasp of the difficulties he faced. Before setting up any new experiment, he would have to sit down and think even harder about his methodology. He needed to pose his question in such a way that whatever result he obtained gave him a clear-cut answer. To take it still further he would need to overcome key limitations of the facilities then available to him at the marine laboratory.

Critically, the marine lab could offer him little or no molecular biological backup. Although he was well aware of the need for genetic confirmation of his work, Williamson could anticipate no subtleties of chemical or genetic analysis here on the Isle of Man. This meant that his initial experiments had to be couched in basic terms, much like those of the Victorian era of observational biology. Only if he could first show, objectively and obviously, that he was on the right track, could he hope to enlist the aid of other scientists to complete the essential genetic and molecular analyses.

If the marine lab had its weaknesses, it also had its strengths. Immediately to hand were huge briny tanks in which scallops, lobsters, and various types of flatfish and round fish were routinely harvested. For a hundred years biologists working here had taken a special interest in the production and study of larvae, some of which are very difficult to rear in captivity. This practical foundation, drawing on his own extensive personal experience of larval biology, was the logical place for Williamson to turn. After giving it some more thought, he focused on an animal that has long fascinated the world of biology—the humble sea squirt.

Sea squirts, which are members of the phylum of tunicates, or ascidians, are regarded as distant cousins of all vertebrate animals, though you might be forgiven for not seeing the resemblance at first sight. They have no heads, eyes, or even limbs. They are often portrayed as hollow flasks, or worse still, mere bags of flesh, squatting unadventurously on the marine substratum, yet for the biologists who study them their appearance belies a subtle complexity and beauty. I well remember my own sense of surprise at first observing the common sea squirt, *Ciona intestinalis*, transilluminated by light. I was enthralled by what resembled the most exquisite narrow-bodied teapot, with walls of gossamer-thin semi-lucent crystal, and gilded around its delicately fluted rim and spout. It took a second hard stare to recognize I was looking at a living creature. Sea squirts are members of phylum of the urochordates—Greek for chordate with a tail—and the adults metamorphose from a tadpole larva. It is in this minuscule but remarkable larval creature, which ranges in size from a mere visible speck to half an inch, that we find the telltale link to the vertebrates. However much they resemble the tadpoles of frogs and toads, the larvae of sea squirts are not directly related to amphibians. Nevertheless, sea squirt tadpoles do have the equivalent of a very primitive brain, a nerve cord, and a heart, and, like amphibian tadpoles, they swim in elegant fashion through undulations of their muscular tails.

Adult sea squirts grow up to a foot in length, some living singly while others form colonies that swarm over hard surfaces, such as tropical reefs, rocky bottoms, and shipwrecks. As a snorkeling swimmer, you might glimpse the mouth—the teapot rim—opening and closing. Take a closer look, and you might spot a clear reminder of its vertebrate linkages: the flash within of eight white teeth. Cilia around the rim set up a current that sucks phytoplankton, as food, through into the stomach, and the waste is squirted back out through the spoutlike second opening.

Given the adult form and lifestyle, it will come as no surprise that before biologists ever linked them to their tadpole larvae, sea squirts were wrongly classified—as mollusks and even as marine worms. Only when, in 1866, the Russian biologist Alexander Kovalevsky recognized that this "gilt-rimmed" teapot developed through a tadpole larva did biologists realize that they were looking at a chordate. For Darwin, it was a discovery to be celebrated, for "it seems that we have at last gained a clue to the source whence the Vertebrata were derived."[1] Kovalevsky was also the first to observe the dramatic, if brutal, metamorphosis, in which the more obvious chordate structures of the tadpole are sacrificed during the transformation to the bottom-fixated adult in a powerful melodrama enacted, in the depths of the oceans, on a Lilliputian stage.

Unlike the starfish, with its extraordinary metamorphosis from bilateral to radial symmetry, the sea squirt is bilaterally symmetrical throughout its life cycle. Logically, we would expect that the adult sea squirt would develop in a simple linear fashion from the basic scaffold of the larval structures, with the adult body plan following the larval axis, and growing, however extravagantly, out of the tissues and organs of the larva. And, indeed, at first all seems to be going in that direction as the embryo develops to the primal stage of the simple ball of cells. But from this point onward, in the words of Professor E. J. W. Barrington, who studied the metamorphosis with undisguised astonishment, "two independent developmental mechanisms are now operating side-by-side, the development of the larval structures proceeding virtually independently of that of the permanent ascidian organisation."[2] Ignoring the larval body shape, and independent even of its essential body axis, the adult grows inside the larval tissues, like an alien invader, a parasite devoted to its own selfish ends.

To begin with, the tadpole is a little ahead, but the adult soon catches up. Even the larval nervous system, including the simple brain—which might be expected to develop into the adult nervous system—is ignored, the two separate brains and nervous systems coexisting, side by side, or one above the other, all in the one tiny body. Soon—it may last anything from a few hours to a few days—the tadpole abandons all pretense of an independent existence and heads for the very different ecology of the ocean floor. Here it discovers a suitable flat surface and glues its head to it, in anticipation of the forthcoming sacrifice. Within minutes, the beautiful tail is destroyed from within, the skin separating from the internal tissues, its fins jettisoned. The notochord is torn apart, the covering sheath ruptured, and the contents consumed by phagocytes. Even the tadpole brain is attacked and devoured, though its annihilation takes longer, days rather than minutes. Meanwhile the skin cells of the tail are transformed and recycled so that all that remains is the emergent form of the adult sea squirt, ready to begin its sedentary existence on the ocean floor.

Like the example of *Luidia sarsi*, this extraordinary development prompts important philosophical questions. This metamorphosis is very different from, say, that of the tadpole and frog, where the frog develops out of the basic form of the tadpole, however much is sacrificed. The adult sea squirt completely ignores the body structures and development of the tadpole larva. How, in such curious circumstances, do we define the meaning of individual life—or death? Should we regard the tadpole as a life-form in its own right, which is ultimately sacrificed? Other species, known as larvaceans, which are included in the same phylum as the sea squirts, exist as free-swimming planktic tadpoles in their adult forms. However much we attempt to rationalize it from a philosophical standpoint, the implications jar—they seem alien to our deep-felt human instincts. Judging from our human experience and sensitivities, this thrilling yet aesthetically disturbing cycle of birth, sacrifice, and rebirth challenges our essential concept of being. Indeed, there are enigmas enough for the most prosaically minded scientist to chew over in the evolutionary and developmental origins of what we are witnessing. It is little wonder that this metamorphosis remains today, as it has been for more than a century, the focus of enormous scientific scrutiny, the sea squirt genome one of the first to be unraveled, its developmental pathways dissected and spread over thousands of computer screens, as twenty-first-century science continues to explore how this humble tadpole larva, presumed by many biologists to be close to the hypothetical ancestor of all chordates, came, in the words of Meinertzhagen and Okamura, to be the bearer of our very own "chordate brain in miniature."[3]

Sea urchins, with their marvelous, if equally brutal, metamorphosis, are no less challenging than sea squirts. Urchin eggs hatch into the planktic waters as ciliated balls of cells, and these, in turn, develop into tiny prism-shaped bodies before extruding eight arms that are internally stiffened with a rod-shaped skeleton and covered in cilia to enable locomotion. These are known as pluteus larvae, so-called because the German naturalist who first described them, Johannes Müller, thought they resembled an artist's easel. The adult sea urchin also develops, alien fashion, from pluripotent stem cells in the abdominal cavity of the pluteus. As with the starfish and the sea squirt, the adult and larval sea urchin are different life-forms, with their own body shapes, means of locomotion, and feeding, and a programmed sacrifice of the larva for the benefit of the adult. For Williamson, the marked differences between the two great marine metamorphoses—the sea squirt, with its tailed tadpole larva, and the sea urchin, with its pluteus larva—set his mind on a new and challenging experiment, one that offered an extreme test of his theory. He would attempt to fertilize a maternal sea squirt, from the phylum of the urochordates, with a paternal sea urchin, from the phylum of the echinoderms.

Plentiful supplies of the sea squirt, *Ascidia mentula*, were being cultured in one of the seawater storage tanks at the lab. This species, shaped like a fat sausage some four or five inches long, would provide the eggs. In 1985 Williamson enlisted divers, wearing aqualungs, to enter the tank, where they scraped the sea squirts off the walls and floor, providing him with his maternal species. For the paternal species, he chose the familiar urchin, *Echinus esculentus*, enlisting the same divers to wade into the bay of Port Erin, where they collected dozens of animals from a ruined breakwater just in front of the marine laboratory.

As before, Williamson milked the sea squirts of their eggs by holding them over a bowl and squeezing their bodies. There was, however, a predictable complication: sea squirts are hermaphrodite. If he was unlucky he collected a cloud of sperm, if lucky, hundreds of eggs. Sometimes he would get both in his collecting bowl, though thankfully this was rare. The eggs were visible to the naked eye as white spots, and he readily collected them with the aid of a magnifying glass. The hermaphrodite sea squirt was also capable of self-fertilization, so he had to observe the eggs over several hours. If they started to divide—the first evidence of prior fertilization—he discarded them. With a nylon mesh, he filtered out those eggs that did not develop, ending up with a couple of hundred. He put these into an observation dish, poured urchin sperm over them, and waited to see what would happen. As he subsequently recalled to me: "It was beyond my wildest dreams that I would ever get a pluteus larva, but I had hopes, at the very least, of getting a ciliated larva."

The experiment did not work the first time. But after four or five attempts at cross-fertilization, some of the hybridized eggs started dividing. He could make this out under a low-power microscope, needing no more than ten times magnification. In most cases the eggs divided a few times and then stopped. This was disappointing, but it was enough to encourage him to try again.

Two years passed as he labored at a series of cross-phyletic hybridization experiments. He grew better through practice, in effect serving his hybridization apprenticeship. By 1987, one of the experiments resulted in two eggs hatching out into spherical larvae. This was an exciting observation, since sea squirt eggs should normally hatch as tadpole larvae. But these hybrid hatchlings looked more like the hatchlings of sea urchin eggs, which emerge as radially symmetrical hollow ciliated spheres. He watched the larvae develop, observing how they lost their radial symmetry to become prism-shaped, with its implication of bilaterian symmetry, and then further, when they sprouted arms. This was thrilling to observe, since it was exactly what he would have anticipated during the pluteal larval development of a sea urchin. But these larvae had come from the eggs of a sea squirt. Williamson was convinced that he had transferred the pluteus larva of a sea urchin into the metamorphic life cycle of a sea squirt—he had moved a developmental program from one phylum to another. Yet standard practice demanded that he repeat the experiment.

The sea squirt normally breeds throughout the year, but the urchin only breeds in late winter and early spring, thus limiting his opportunities to the urchin breeding season. In 1988 conditions proved right, and he repeated the experiment. He was delighted to witness exactly the same outcome. "The larvae were very like echinus larvae but developed from ascidian eggs."

The following year, he conducted another experiment, with much the same result. In all these experiments, the larvae developed to the pluteus stage but failed to develop any further. By then they were about a month old. For a cross-phyletic hybridization experiment, it was a remarkable achievement. A few years earlier, Williamson had first proposed that larval transfer might have taken place in nature through hybridization between different species. At the very least his experiments appeared to show that it was possible. Anyone who has watched marine wildlife programs on television will have been impressed by the hit-and-miss nature of broadcast spawning, when vast quantities of eggs laid by the females take their chances with clouds of sperm from the males, as they float and drift in the surface waters. In such circumstances, the sperm of one species would have ample opportunity to come into contact with the eggs of another. Williamson had been successful in just a few hugely optimistic experiments. What opportunities nature would have had, over the vast time periods of evolution, for countless such experiments! His theory required only an occasional success with reproductively viable offspring to bring together the genetic programming of two different species into a single life cycle.

For more than a hundred years, all explanations of marine metamorphosis, no matter how varied, had played, like musical variations, around the same elementary motif. Evolution worked slowly, along the straight lines of linear Darwinian descent, with natural selection operating on changes based on mutation—which itself presupposed that change arose strictly within individuals through copying errors in the genes when cells divided. Now, if they heeded Williamson, biologists were offered the opportunity of thought and debate at a more fundamental level—the potential for relatively sudden change, change that came from the sexual merging of pre-evolved and very different genomes. Such a broader perspective should stimulate science. But Williamson did not kid himself that the conservative world of evolutionary biology would welcome a radically different interpretation of common assumptions. Nevertheless, he felt ready to publish his theory and the results of his experiments, in the hope that they would be considered without prejudice.

— 4 —

The Price of Iconoclasm

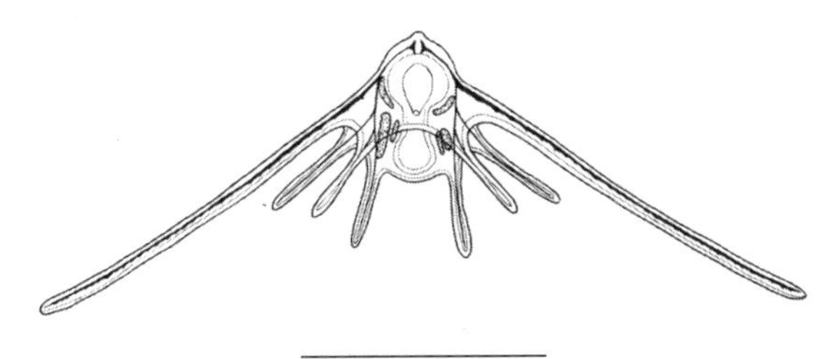

Pluteus larva of a brittle star

In June 1985, Williamson wrote out his theory of larval transfer in the form of a scientific paper. The title was "Incongruous Larvae and the Origin of Some Invertebrate Life-Histories." When he submitted it to the *Biological Journal of the Linnean Society*, one reviewer favored publication while another opposed it. The editor refused full publication but offered to publish a short note, summarizing Williamson's ideas. Williamson declined. In October he submitted the same paper to *The Journal of Natural History*, and once again one referee was in favor and two against publication. In the midst of these rejections, Williamson came across an intriguing article that the marine biologist H. Barraclough Fell had written many years earlier.[1] An internationally acknowledged expert on the echinoderms, Fell currently held a chair at Harvard; but in 1941, when he had written the article, he was working in a more junior capacity in his native New Zealand. In the article Fell told of an unusual discovery that his former colleague and professor, H. B. Kirk, had made some twenty-five years earlier—a discovery that, for some obscure reasons, had never previously been described in detail.

Kirk regularly perambulated the rocky coastline of a small headland that separates Island Bay and Ohiro Bay, at Wellington, on the northern shore of New Zealand's Cook Straits, a site well known for southerly gales. One day in 1916, during a period of extremely low spring tides, he came across clusters of fertilized eggs attached to some stones. Kirk had no idea at the time what marine creature had laid them, but he brought them back to his laboratory and monitored their development. He and his colleague, Fell, discovered similar clusters in the same spot on a number of occasions over the ensuing twenty-two years, thereby establishing that the annual spawning season for the mystery species was in the latter half of August, the eggs being laid during the high tide but only becoming visible at low tide.

That first year, 1916, Kirk recorded his initial observations: "The eggs are spherical, 0.5 mm in diameter, each with a perfectly transparent, thin, but extremely tough chitinous envelope." So tough was this covering that Kirk could not remove it without crushing the embryos. And in spite of the envelope's transparency, he could not discern what was really going on inside the embryo because the embryo itself was opaque. His only recourse was to sacrifice some of the embryos and dissect them, but this would have to keep until the following year. In the meantime, he was content to watch and record whatever developed.

To his astonishment, the eggs hatched out not as larvae but as junior adults, a development known as "oviparity." More astonishing still, the junior adults matured to an unknown species of brittle star, a separate class of echinoderm closely related to starfish, and easily recognizable because of their elongated whiplike arms, which they use for locomotion over the ocean bottom.[2] Kirk wrote up his discovery in a preliminary paper in which he recorded the measurements of the eggs and the details of the hatching process, together with brief notes on this puzzling species.[3] At first he thought he had found an unusual example of the Australian brittle star, *Ophionereis schayeri*, but this species metamorphoses through a non-feeding larva, whereas his discovery had no larva. Over time he realized that the development of this mysterious brittle star, from hatchling to adult, was "absolutely direct," with "no evidence whatsoever" of a metamorphosis through a larval stage. As far as I am aware, the species of brittle star is still unattributed, but has come to called "Kirk's brittle star," or as Williamson labels it, "Kirk's ophiuromorph."

When Fell, in his turn, reared specimens from eggs in small dishes through which seawater was slowly circulated, he confirmed, in finer detail, that the development was markedly different from what he would have anticipated in a brittle star. To understand what he was observing, indeed to grasp its importance, we need to know a little more about the development of the echinoderms.

Until recently, the phylum of the echinoderms consisted of five classes: the sea urchins, the starfish (marine biologists prefer to call them sea stars, since they are not fish), the brittle stars, the combined feather stars and sea lilies, and the sea cucumbers. Recently a sixth class, known as the sea daisies, has been added—though some biologists think they should be included with the starfish. Sea daisies have no larvae, and they are radially symmetrical throughout their development. Of the species in the five remaining classes of echinoderms that do undergo metamorphosis, the adult forms are always radially symmetrical, yet the larvae are always bilaterian. For biologists who assume a linear evolutionary history, this dramatic change of symmetry within a single life cycle poses a dilemma. Given that the internal organization, and thus the genetically programmed development, of a radial creature is very different from that of a bilaterian creature, how, as part of a linear descent-with-modification evolution, did the genetic blueprint and structural organization of a bilaterian larva evolve into that of a radially symmetrical adult?

In fact the simple truth is that, at the time Williamson was formulating his hypothesis of larval transfer, there was no generally accepted explanation for this dilemma. It was assumed that, based on the larval symmetry, the echinoderms had evolved from simpler animals with bilaterian symmetry. Somehow, and it was further assumed this would have come about through mutation and adaptation through natural selection, the phylum had evolved a radically different radial symmetry. This radial symmetry was not completely spherical. The echinoderms have an upper and lower surface. Rather it is radial in the horizontal plane. Or, to be more precise, it is pentaradial—divided in the horizontal plane into segments, rather like an orange, based on a rule of five, so there might be five, or multiples of five, segments.

How then, given the lack of orthodox explanation, can we attempt to examine this dilemma further? One avenue we might reasonably explore is the mechanics of the metamorphosis—and, to put it mildly, this is bizarre.

We have already seen, in the example of the starfish *Luidia sarsi*, how in echinoderms with larval development, the juvenile adult begins as a cluster of cells within the larval body. We witness two distinct developmental programs arising from the same original fertilized egg and taking place within the tissues of a single life-form. The metamorphosis is accompanied by massive internal change coupled with catastrophic destruction of the larval tissues. Huge chunks of the larval body, its tissues and organs, are digested away and reabsorbed, or simply discarded. There are similarities in this "catastrophic" form of metamorphosis to what is seen in the pupal stages of certain orders of insects, a topic we shall return to later. Williamson asks, not unreasonably, how, such dramatic changes in symmetry and such dramatic reorganization, involving wholesale destruction and waste, could have evolved through any linear process of evolution.

To dig deeper into this riddle, and to link it with Fell's observations on Kirk's brittle star, we need to know a little about embryonic development, and how this links to the basic branches of the tree of life.

The fertilized egg is spherical, especially so when supported by the buoyancy of oceanic water. Thus animal life begins with radial symmetry. Fertilization triggers the process of cell division within the egg, leading to the formation of a hollow ball of cells, known as the "blastula," which in echinoderms is how they usually hatch from the egg into the ocean, their minuscule bodies ciliated to allow locomotion. In the next step of development an organized sequence of cell movements within the blastula leads to the formation of three germ layers. The outer layer, or "ectoderm," will form the skin and nervous tissue; the inner layer, or "endoderm," will form the lining of the gut and internal organs; and the cells in the intermediate layer, or "mesoderm," will give rise to internal tissues and organs such as muscle, bone, and heart. A dimple now appears on the surface of the blastula, and cells tuck in to form a tunnel leading into the interior. The dimple becomes the "blastopore" and the tunnel itself becomes the primitive gut, marking the change from blastula to "gastrula." What happens next sets the stage for the divergence of the two fundamental trunks or pathways of animal evolution at the base of the animal tree of life.

For those member species of one great trunk, henceforth known as "protostomes" (from *proto* for "first" and *stoma* for "mouth"), the blastopore develops into the mouth. This protostomal lineage gave rise to the present-day mollusks, marine worms, insects, spiders, and crustaceans such as crabs and lobsters. For the member species of the other great trunk, the blastopore develops into the anus. In these species, which includes our own, the tubelike gut breaks through to make a second opening onto the surface, and it is this second opening, and not the blastopore, that develops into the mouth. These species are known as "deuterostomes" (from *deutero* for "second" and *stoma* for "mouth"), and they include the chordates and thus the mammals.

These two lineages, which date back to the earliest origins of the animal kingdom, embrace a puzzle when it comes to the echinoderms. Where the great majority of the marine invertebrates belong to the protostomes, the echinoderms—those most colorful and bizarre of the marine invertebrates—are classed as deuterostomes. This placing of the echinoderms in the deuterostome trunk is entirely due to their larval development. And in this respect Kirk's brittle star appears anomalous.

Just like our own human, vertebrate, embryo the embryo of an echinoderm develops as a deuterostome. But when it came to Kirk's brittle star, Fell observed that its development did not follow a deuterostome pattern. The fertilized egg developed to a radially symmetrical blastula and this, as usual, developed to a radially symmetrical gastrula. The initial tunnel leading from the blastopore was unusually small, and the primitive gut that extended in from it soon filled up with cells. The blastopore itself, though its position could be seen as a dimple on the surface, now led nowhere. After six or seven days, five grooves radiated out from this blind dimple, and the embryo became radial, with five segments, like those of an orange, separated by grooves running toward the midpoints of the sides. This species of echinoderm appeared to be radially symmetrical throughout its development. Over the next ten days, Fell observed a novel development, with the adult brittle star forming directly from this pentaradial embryo. During this time the blastopore reestablished a tunnel to the interior, which developed into the stomach and intestine. There was no second opening, no deuterostomal mouth.

In Fell's words, "The blastopore occupies the position of the future mouth. This is, of course, contrary to the usual fate of the blastopore in [brittle stars] since it normally gives rise to the anus of the larva."

Fell arrived at two possible explanations for what he was observing. One was that the ancestral stock of this strange new brittle star had always developed directly. But if this was the case, direct development in this species of brittle star was both pentaradial and protostomal. The other possibility was that his brittle star had evolved from an ancestral species that metamorphosed in the normal manner of brittle stars, but subsequently the larval stage had been suppressed. Larvae were known to be suppressed by cold climates, and he wondered if this putative ancestor had lost its larva in response to the severity of the most recent ice age.

There may be a hint of larval suppression in the vestigial tunnel leading from the blastopore in the early gastrula. But if Kirk's brittle star really did evolve from an ancestor that metamorphosed, the suppression of the larval phase would appear to be unusually complete. And there was no escaping the fact that this first dimple, the blastopore, became the mouth. Williamson rejected any explanation based on larval suppression, pointing to the evidence of Fell's own findings. Where echinoderms with larvae should develop along deuterostome lines, and with a bilaterian larva metamorphosing to a pentaradial adult, this brittle star, without a larva, appears to be developing in an exclusively pentaradial manner along classical protostomal lines. Kirk's brittle star appears to be a protostome, which would place it on the opposite branch of the evolutionary divide.

An additional observation adds weight to Williamson's interpretation. The deuterostome versus protostome distinction is often accompanied by another important difference in development. In deuterostomes, the internal body cavity, or "coelom"—what we call the abdominal or peritoneal cavity in humans—usually forms by budding off from the ancient gut. Imagine a blind sac budding out from the gut wall, ballooning inward and gradually expanding and swelling, until it fills the internal space, meanwhile folding over the internal organs to form a covering layer. In protostomes, the coelom doesn't grow out of the gut but forms out of a split in the tissue known as mesoderm. This pattern is known as "schizocoely," from the Greek *schizo* for "split" and *coelom* for "cavity." In Kirk's brittle star, the body cavity developed from this split in the mesoderm rather than from budding of the gut wall. This meant that Kirk's brittle star was not merely a protostome but a so-called schizocoelous protostome. Although this might seem obscure to a lay reader, it is of fundamental importance to evolutionary and developmental biologists. The division of the animal kingdom into deuterostomes, which develop through gut budding, and proterostomes, which develop through mesoderm splitting, is the basis of the primal bifurcation of the tree of life.

How then had Williamson's more orthodox colleagues come to terms with Fell's incongruous discoveries, which had, after all, entered the scientific literature almost half a century earlier?

When Williamson studied the literature he confirmed that biologists had indeed been startled by this curious report of brittle star development. Inevitably, many, such as L. H. Hyman, had favored Fell's second hypothesis, believing that Kirk's new species must represent an extreme case of suppression of the larval phase.[4] But this ignored the schizocoelous inference. Moreover, where larval suppression was a familiar finding with some echinoderms, in no other case did this lead to the complete abandonment of the deuterostome lineage. For both these reasons Williamson thought larval suppression an implausible explanation. In Fell's paper, Williamson also came across another example of direct development in a brittle star. As long ago as 1891, an Italian biologist, A. Russo, had described a protostomal development in another brittle star, *Amphipholis squamata*.[5] Russo's findings had been dismissed as "improbable in the highest degree."[6] In 1968, some twenty-seven years after his original paper, Fell observed that brittle stars of two other families probably developed in a similar manner to Kirk's species.[7] But nobody appeared to have followed up Fell's observation. Williamson subsequently discovered two more examples of the same anomaly. The eggs of the sea daisy, *Xyloplax medusiformis*, develop directly within the ovary and never show bilaterian features, although the fate of the blastopore, whether anal or oral, is currently unknown;[8] and *Abatus cordatus*, a subantarctic heart urchin, broods its eggs, giving birth to live young, and it also develops as a protostome and schizocoel.[9]

To Williamson's thinking, Kirk's brittle star was the exception that proved the rule. According to his theory, the echinoderms had originally developed to full adulthood as radially symmetrical animals without ever going through a bilaterian larval phase. But before discovering Fell's paper, he had assumed that all such directly developing echinoderms were long extinct. Now it seemed that he had been mistaken. And this in turn reinforced his view that the evolutionary origins of the phylum of the echinoderms were different from what was commonly assumed.

Adult sea urchins bore scant resemblance to adult brittle stars.

Yet both classes metamorphosed through similar pluteus larvae. The echinoderms that metamorphosed through larvae were classed as deuterostomes, but, as he now proposed, these anomalous brittle stars and heart urchins, which developed without metamorphosis, should really be classed as protostomes. How could this be reconciled with orthodox theory, which proposed that both classes of echinoderms shared a common ancestor from which they inherited their developmental pathways? In analyzing his own findings, Professor Fell could hardly miss those same anomalies. Indeed, he concluded that any close relationship between sea urchins and brittle stars, as inferred by larval features, was now too preposterous to warrant serious consideration.

It seemed to Williamson that his larval transfer theory might explain the long-standing enigma of how a bilaterian larva metamorphoses to a pentaradial adult. Where most textbooks of invertebrate zoology proposed that echinoderms evolved from bilaterian ancestors, Williamson now believed that the ancestors of the entire phylum of the echinoderms had been protostomal and pentaradial throughout their original evolution, only later acquiring the bilaterian larvae in their development. The acquisition of larvae from a bilaterian deuterostome lineage had masked their original evolutionary trajectory and confused the taxonomy of the echinoderms for one and a half centuries. It was a bold and highly controversial interpretation.

Williamson now revised his paper and sent it to the marine ecologist David Cushing, a Fellow of the Royal Society, asking the distinguished scientist if he had any idea where he might get his paper published. Cushing sent copies to two other fellows, after which he returned a complimentary opinion voiced by all three fellows, making constructive editorial suggestions. One of these fellows, a member of the society's editorial board, expressed the opinion that the paper was "too philosophical for any of the Society's journals." It seemed ironic to Williamson that these journals included *Philosophical Transactions of the Royal Society.* The fellow suggested that Williamson consider submitting his paper to *Biological Reviews.* In May, undaunted, Williamson submitted the paper to the *Oxford Surveys in Evolutionary Biology*, a journal with the specific aim of publishing "new ideas on evolution and to promote controversy." It was once more rejected by the editor, who commented, "It is controversial . . . not really the sort of paper we had in mind." That September, and again the following March, Williamson's paper was rejected in turn by the *Journal of Natural History*, *Biological Reviews*, and *The Journal of Zoology.*

Science, of course, is rightly conservative and adheres to principles that have withstood the test of time. But the publication of a new theory does not assume it has been cast in stone—merely that it has sufficient credibility to be considered on its merits by a wide and diverse readership. Despite two years of persistent rejection of his paper, Williamson had continued to break new ground in his experiments, and was increasingly convinced that hybridization was an important source of evolutionary innovation. Why then was he experiencing this wall of rejection?

There are a number of possible explanations, including differences of personal opinion and simple incredulity, but over and above all such considerations looms an explanation that seems overwhelmingly likely. These are Williamson's own words: "When Darwin published his *Origin of Species* in 1859 it was considered heretical by most people. Today the situation is reversed, and, in biological circles at least, it is considered heretical not to agree with Darwin."[10]

What then was the "heresy" Williamson was perpetrating?

Williamson is hardly challenging the basis of Darwinian evolution. There can be no doubt that he is also a serious and reputable biologist. At the time he was fruitlessly submitting his paper, he had published more than seventy original scientific studies covering research into marine biology and planktic larvae in particular. He remains a distinguished world authority in his field. It is tempting to assume that no heresy appears minor to a true believer, but that would in turn be unfair to the journals involved and to the integrity of the referees who had turned down the opportunity of publication. The simple fact is that Williamson's theory challenged one of the central pillars of biological orthodoxy. This pillar, which is the very basis of the evolutionary tree of life, is known as the biological species concept, or BSC.

Most people feel a sense of awe at the diversity of nature. By "diversity" we mean that there are a vast number of different lifeforms, each belonging to its own group, or species. Any individual species is regarded as distinct from other species, to the extent that individuals from one species typically cannot sexually reproduce with another. For example, it is perfectly obvious that a bird cannot cross-reproduce with an earthworm or an elephant with a housefly. Biologists call this "reproductive isolation." And reproductive isolation has been regarded, since the time of Darwin, as a key aspect in the origin of new species. But in cases of similar species belonging to a common group, or genus, the degree of distinctness, or reproductive isolation, may not always be as clear as this. Appearances are not everything. At one time a yellow-rumped warbler species that inhabited the west of America was known as Audubon's warbler. This bird has a yellow throat that distinguishes it from the eastern warbler, known as a myrtle warbler, which has a white throat. Then it was discovered that the two supposedly distinct species interbred with each other over a wide geographic territory stretching from Alaska to southern Alberta, so that, today, the Audubon's warbler and the myrtle warbler are regarded as subgroups within the same species rather than separate species.

By and large, species do keep to themselves, and thus species tend to follow their own linear trajectories in their evolution. Not only do different species avoid crossbreeding with each other, there are also important physical and genetic differences between them that help to preclude this. While there is an active debate on which takes priority, the reproductive isolation leading to genetic differences, or the genetic differences leading to reproductive isolation, the concept of distinct and separate species, exceptions such as Audubon's warbler aside, is fundamental not only to the Linnean classification of life but also to the core concepts of neo-Darwinian evolution. Modern Darwinians will often use the term *speciation* as synonymous with the mechanism of origin of new species, implying that this is the exclusive means of generating new species from old. Any theory that proposes the crossover of genes, including developmental control genes, or, most spectacularly of all, whole genomes, between different species, challenges these core beliefs. It is hardly surprising that Williamson was experiencing difficulty in convincing orthodox reviewers of the credibility of his theory.

On the other hand, science, by definition, is the pursuit of knowledge, and for knowledge to progress, scientists, however skeptical, must be prepared to consider new ideas. In 1987, and quite out of the blue, Martin Angel, the editor of a journal called *Progress in Oceanography*, contacted Williamson and expressed an interest in his theory. Angel had heard Williamson talk on his larval transfer theory to the Challenger Society for the Promotion of Oceanography, and he had seen an early draft of the paper. As a fellow planktologist, Angel sympathized with Williamson. He frequently came across anomalous larval forms himself and he was puzzled by the taxonomic relationships being put forward by colleagues who refused to question conventional explanations. Angel also knew about Williamson's difficulties with the other journals. He declared flatly, "I'll publish your theory in my journal."

— 5 —

Challenging the Tree of Life

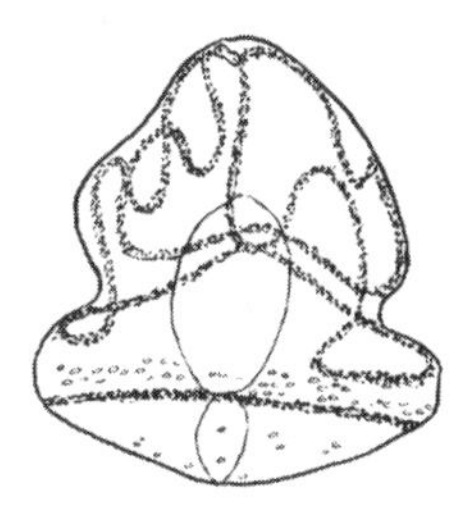

Tornaria larva of an acorn worm

When he published Donald Williamson's paper that same year, Angel introduced it with what must be one of the most unusual editorial comments ever included in a modern scientific journal:

> It is sometimes stated that papers involving ideas are more difficult to get published than those that follow the conventional wisdom of the time. Iconoclasts are rarely popular with their peers. Here is a paper that I am including in *Progress in Oceanography* as something of an editorial indulgence, but I am confident that in time the decision will be vindicated.[1]

Angel went on to list all of the paper's refusals and rejections. From the content of some of the reviewers' comments, he was convinced that prejudice had blinded them to the logic of Williamson's arguments. "Darwin," he wrote, "would have probably had less trouble submitting a draft of the *Origin of Species* to the Bishop of Oxford."

With Angel's help, Williamson was at last in a position to tell his colleagues why he disagreed with the prevailing view of animal life history, in which larvae were assumed to fit the same linear evolutionary development as their corresponding adults. It was a risky if courageous undertaking.

Over the century and a half since Darwin introduced his theory of evolution, leading authorities from many countries had used larval features as their guide to fitting entire classes of life into positions in the evolutionary tree. Such relationships were typically depicted as a linear development, in which new life-forms exclusively arose through branching from preexisting species. The implications for larvae and the evolution of metamorphosis, were clear: the various stages of development, from fertilized egg, to embryo, through one or more larval stages and on to the adult life-form, could only have evolved as part of this linear "descent-with-modification" process.

This "linear-with-branching" explanation had an understandable logical appeal. Moreover, as Darwin himself had pointed out, it allowed for the evolution of considerable variety at the larval as well as the adult stage since each stage of the metamorphosis was itself subject to natural selection. In most of the commonly described life histories, the larval form was an adaptation to its free-floating planktic ecology just as the adult form was an adaptation to its more sedentary ocean-floor ecology. Sometimes the adults showed more evolutionary experiment than larvae; in other cases the larvae were the main experimenters. But in one respect all were agreed: metamorphosis through its various stages was so radical a development that there had to be very powerful evolutionary advantages for its common occurrence and persistence.

The obvious and overriding advantage for a larval stage was mobility. Surface currents swept those "wanderers" far and wide, vastly extending the geographic and ecological range of the species. All of this Williamson accepted. But it hardly explained why creatures from the same genus, family, class, or phylum, whose larvae occupied the same ecology, should sometimes have very different larvae. No more did it explain why the same larval forms cropped up in the life history of animals from different genera, families, classes, or phyla, with no intermediate forms bridging the evolutionary gap, any more than it explained, to his satisfaction, why some life-forms have two or more quite different larval phases, with all the waste, danger, and expended energy this entailed, while inhabiting the same planktic ecology. In his paper, Williamson suggested that all such puzzles could readily be explained if large amounts of genetic material had been transferred between different species at some stage in their evolution. Take, for example, the anomaly of the sponge crabs, which have a very similar larva to the very distantly related hermit crabs: this made perfect sense if the earliest sponge crabs had no larval phase but acquired this later from a hermit crab by genetic transfer.

If Williamson was right, not only would his theory explain the many anomalies he had seen in the tree of life, it would also explain how echinoderms' bilaterian larvae developed into radially symmetrical adults. The ancestral echinoderms had never evolved as bilaterian, but were radially symmetrical throughout their evolution. This would explain the development of Kirk's brittle star and a number of other supposedly anomalous developments among the echinoderms. In a major evolutionary event at some subsequent time, though still in the very distant past, echinoderms had, after they had first evolved into the major classes, such as the starfish and the sea urchins, acquired the genetic template for their bilaterian larvae from another species.

Darwin shared Haeckel's view that the development of animals from egg to adult holds irrefutable evidence of their past evolutionary pathway. Again and again, Darwin emphasized its importance as a key to the most profound level of understanding. "On this view," he wrote, "we can understand how it is that, in the eyes of most naturalists, the structure of the embryo is even more important for classification than that of the adult."[2]

Today we regard metamorphosis as a continuation of embryonic development outside the egg, embracing its startling changes within the umbrella of "post-embryonic development" or "reprogramming." This was a view that Darwin himself anticipated. Not only did he devote a major section of *The Origin of Species* to a detailed discussion of embryonic development, including metamorphosis in insects and marine animals, he explicitly acknowledged that an understanding of metamorphosis was "one of the most important subjects in the whole round of natural history."[3] Well aware that the striking changes seen in metamorphosis might weaken his argument for a slow and linear change under the influence of natural selection, he downplayed their challenge, declaring that the changes seen in the more dramatic examples of metamorphosis, for example, from the caterpillar to the butterfly, were unrepresentative. "The metamorphoses of insects, with which every one is familiar, are generally effected abruptly by a few stages; but the transformations are in reality numerous and gradual, though concealed."[4]

To grasp Darwin's point, we might look to the example of a secretive group of creatures, acorn worms, whose ecological importance is easily overlooked.

The acorn worms inhabit U-shaped tunnels in the seabed, constructing tubes of mud or sand particles around themselves, and, much as earthworms on dry land, vacuuming organic material from the sand debris of the world's shorelines. While the casual beach walker might not be familiar with the worms themselves, he or she could hardly miss the characteristic coils of sand discarded from the anal end of the tube. Acorn worms also resemble earthworms in shape but not in size—some of them grow to eight feet in length. And their internal structures also set them far apart from earthworms on the tree of life. Behind the proboscis is a cylindrical collar that conceals the mouth, and, in the main body behind the collar, are gill slits opening into the throat cavity. The features of the acorn worms, together with their two nerve cords, place them as a separate class within the phylum of hemichordates, which, together with the urochordates and our own phylum of the craniates, make up the superphylum of the chordates.

The acorn worm larva is known as a "tornaria" because, in shape and the blur of its rotatory movement, which is readily visible under a dissecting microscope, it looks like a child's spinning top.

In its subsequent metamorphoses, this larva undergoes considerable changes in shape and internal structure to become the greatly elongated and burrowing wormlike adult. However, all these changes are gradual and linear, much as Darwin envisaged. The front part of the larva elongates into a proboscis, then a collar forms behind this, and the remainder of the larval body elongates into the multigilled trunk. The respective parts of the juvenile body develop from the corresponding larval structures, and the only larval features to be discarded are the bands of locomotory cilia. For Williamson, at the time of drafting his first paper, it was reasonable to suggest that a linear process of evolutionary change involving descent with adaptation might well explain the sea acorn metamorphosis.

Darwin's linear and gradualist explanation had been accepted and endorsed by generations of evolutionary biologists. But now, in putting forward a theory of evolutionary change that involved transfer of whole genetic programming between dissimilar species, Williamson was contradicting such orthodoxy. The introduction of a novel developmental program into an existing species would bring about an abrupt and major evolutionary change—what biologists call a saltation. The evolutionary consequences could hardly be described in terms of a branching tree. On the contrary, they implied that different branches would recombine, to create, in a single dramatic event, a new life-form.

However, returning to the larval forms of acorn worms and echinoderms, we encounter an anomaly. In 1881, the Russian zoologist Elie Metchnikoff, famous for the discovery of our phagocytic white blood cells, drew attention to the fact that the acorn tornaria also resembles the larvae of certain echinoderms, in particular the starfish. So confusing is the similarity that nineteenth-century biologists mistook the acorn worm larva for that of an unknown family of starfish. During their respective developments from the fertilized egg, the echinoderms and acorn worms share not just the form of the larva, they also share the same process of internal body cavity, or coelom, formation as well as a radial form of early cell division.

So close are these resemblances that biologists have wondered if both the echinoderms and acorn worms might have evolved from a common ancestor, perhaps one that resembled a tornaria. But any such link depends entirely on the larval similarities. Meanwhile, the marked differences between the adult forms are very difficult to reconcile with the larval similarities. This is why H. B. Fell suggested that biologists should ignore the larval developments and consider only the adult developments when working out evolutionary relationships. The distinguished biologist Pat Willmer of the University of St. Andrews agrees with Fell: "Very considerable problems [are] raised by the diversity of echinoderm larval forms. . . . Any classification of echinoderms based on larval form is totally discordant with adult morphology and with fossil evidence." In Willmer's opinion, the similarities between the larvae may be merely convergent and "perhaps even meaningless."[5]

But convergence usually denotes a physical or superficial resemblance, and Williamson points out that in a number of complex internal developments the tornaria really does appear to show the same origins and patterns as, say, the two-ringed "bipinnarial" larva of the starfish or the ear-like "auricularia" larva of the sea cucumber. In his view, this makes convergence unlikely.

Williamson disagrees with both Fell and Willmer, proposing that we should not ignore the larval forms—rather we should draw a different, indeed a radical, even revolutionary, conclusion. Both the adult and the larval forms become illuminating if we accept that the echinoderms and hemichordates evolved independently and only became related when a hybridization event took place between early species of the two phyla, and this in turn gave rise to the common larval forms. This requires no bizarre switch from bilaterian to radial symmetry in the life history of the echinoderms—it assumes that they retained their primal radial symmetry—while the arrival of the bilaterian larvae into the present life history came about through hybrid acquisition. Given the hit-and-miss nature of broadcast spawning in the oceans, the opportunity for hybridization between different species would have arisen very many times. Each putative hybridization event would have led to a specific saltationist genomic experiment. Williamson conjectures that many such genomic experiments would have taken place over the vastness of evolutionary time, but the majority probably ended in failure. All it would take is an occasional successful outcome for the hybrid offspring to possess the genetic programming for two very different body plans.

Any potential offspring would inherit the developmental blueprints for two different life-forms, a simpler one following the planktic life cycle of the larva and a more complex blueprint for the adult existence on the bottom of the ocean. The planktic phase, in such a hypothesis, evolved to the present-day larvae.

To Williamson, it seemed likely, given the close resemblance between the larva of the sea cucumbers and the tornaria of the sea acorns, that hybridization of an early echinoderm with an ancestor of the present-day acorn worms would offer a logical explanation of how the echinoderms acquired their first larva. From this reticulate evolutionary event, over millions of years, the genetic blueprint for this larval form, or for other related larvae, now evolving from it, spread throughout the other classes of echinoderms.

With Angel's help, Williamson's theory was at last presented to his colleagues for due consideration. If he was right, conventional theory of linear descent had so blinkered our vision it had shifted the entire phylum of the echinoderms from their proper position on the evolutionary tree. They were not deuterostomes, like the vertebrates, but members of the opposite great trunk, the protostomes, which included the superphylum of the arthropods, such as lobsters, crabs, shrimps, insects, and spiders. Indeed, if his biological colleagues accepted his line of argument, the elegantly simple tree of life would have to be chopped down at its first great fork and a new, more complex tree would have to be drawn up, with the echinoderms transferred from the deuterostomes to the protostomes, and with some branches and twigs reuniting as well as dividing.

6

"This Is Impossible!"

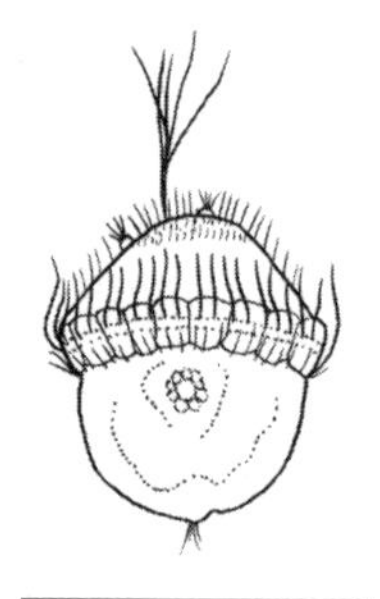

Trochophore larva of a mollusk

In the spring of 1988, Lynn Margulis, Distinguished Professor at the University of Massachusetts Amherst, received an unsolicited reprint of a scientific paper in her mail. Arriving out of the blue, and without an accompanying letter, it outlined a remarkable and completely novel theory about the putative role of larval transfer in marine metamorphosis. It was written by a man she had never heard of, a marine biologist called Donald Williamson working in some obscure laboratory on the Isle of Man, in the far-off Irish Sea. She read the paper and was deeply impressed. She wrote back, "Who are you? I've never heard of you. What a marvelous theory!" Williamson's reply was typically humorous: "There's no reason why you should have heard of me. I'm not an evolutionary biologist. I've been a planktologist all my life, so there's no reason why you should know me. . . . I am sixty-six years old, from a family whose members are short-lived and thus I'm on a straight-line course for posthumous recognition."

In retrospect, Williamson's approach to the Amherst-based scientist would prove to be important in obtaining recognition for his theory. In January 2001, on meeting him for the first time at his home at Port Erin on the Isle of Man, I asked him why he had taken this step at that particular moment in his life.

"I knew," he explained, "that she had a theory about endosymbiosis as the origins of cells, which, in some ways, was similar to my larval theory. So I thought she might be interested."

Today Lynn Margulis is a famous figure in the world of biology. A member of the American National Academy of Sciences, she has been honored with many international awards, including the U.S. National Medal of Science. But in 1970, when her book *Origin of Eukaryotic Cells* was first published, she was a youthful thirty-two, and mother of young children, working as a relatively unknown assistant professor in the Department of Biology at Boston University. Her serial endosymbiosis theory, or SET, proposes that the first eukaryotes—life-forms with nucleated cells—came into being through the symbiotic union of bacteria-like ancestors to create a new composite being, one that ultimately led to the kingdoms of plants and animals as well as fungi and many of the single-celled creatures known as protists.[1] Margulis had endured a protracted battle with scientific orthodoxy before her theory became widely accepted. Now it was Williamson's turn to battle those same conservative forces with a new theory of evolution, based not on symbiotic union but on hybrid sexual crossing. As generous of spirit as she was erudite, Margulis had the wisdom to consider a revolutionary new theory with an open mind, coupled with the strength of will to combat the inertia and antipathy of her more conservative colleagues. In reading Williamson's paper, she could only be fascinated.

In 1988, Williamson submitted a new paper on larval transfer to both the leading science journals in the English language, *Nature* and *Science*. Both journals rejected it. In this paper, Williamson had described how he had fertilized sea squirt eggs with sea urchin sperm, resulting in sea urchin (pluteus) larvae. The referees declared that this was impossible. Margulis suggested he write a shortened paper for the reputable American journal *Proceedings of the*

National Academy of Sciences, or *PNAS*. Williamson now modified his paper to suit the requirements of *PNAS*. The editor sent it out to four reviewers, whose responses proved "inconclusive." More opinions were sought. One reviewer point-blank refused to read the paper, branding Williamson's ideas "high school level rubbish." Another thought the fossil record was consistent with Williamson's ideas, but he was unable to evaluate Williamson's developmental arguments because the territory was highly specialized even for an experienced biologist. A third commented that Margulis should end her support for Williamson forthwith before her scientific reputation was jeopardized. A consistent criticism was the need for better photographic documentation of the hybrid organisms. Margulis concurred. Others thought, not unreasonably, that molecular studies were needed to confirm or refute Williamson's claims. At the same time, nobody could deny that his theory was abundantly clear in its proposals and should be readily testable by experimental confrontation and by standard molecular biological investigation.

While the paper was still undergoing peer review, Margulis introduced Williamson's ideas to Alfred I. (Fred) Tauber, MD, professor of medicine at Boston University School of Medicine. Tauber was also the director of the Center for the Philosophy and History of Science at Boston University and was currently arranging a symposium on the subject of "The Self." When Tauber asked Margulis for names of appropriate speakers, she had no hesitation in recommending Williamson: his theory challenged the very notion of selfhood. In Tauber's words, "Lynn recommended Williamson and the controversy surrounding his fascinating work." But Tauber understandably felt obliged to question what the controversy really amounted to. Was Williamson being "very clever but deceptive?" Or was he proposing one of the most interesting ideas in a century involving marine zoology? Could he convince a sophisticated audience that his theory was based on adequate observation and sound scholarship?

Ahead of the symposium, and during a coincidental trip he had already planned to the United Kingdom, Tauber detoured to the Isle of Man, meeting up with Williamson in the marine laboratory at Port Erin. It is evident, from his subsequent comments, that Tauber's fears were put to rest. In Williamson he believed that he had come across one of the last great nineteenth-century-style naturalists. Writing back to Margulis, he confirmed that the marine biologist was an erudite and careful scientist whose originality merited airing and appropriate criticism. Williamson was well known as far afield as Japan and Korea for his previous work on shrimps and other crustaceans. By then retired from his position at the marine lab, he continued to work in an emeritus capacity, with access to laboratory and office accommodations, but without genetic and molecular biological support. "Although he has little in the way of modern technology, he has performed decades of careful work with equipment at least as good as that of Ernst Haeckel . . . on a theory based solely on observation, intuition, a broad grasp of phylogenetic relationships, and the tantalizing first experiments that appear to support his argument."[2]

Williamson felt encouraged to continue with his work. In 1989, he repeated the cross-phyletic experiment, breeding eggs of the sea squirt, *Ascidia mentula*, with the sperm of the sea urchin, *Echinus esculentus*. The new experiment was spectacularly successful, and he obtained more than 230 fertilized eggs, a great many of which developed into pluteus larvae typical of the paternal sea urchin. What was this if it was not larval transfer? He ran parallel control experiments in which he fertilized sea urchin eggs with sea urchin sperm, and these produced indistinguishable pluteus larvae, which developed at the same rate. With growing excitement, Williamson followed the progress of his test larvae. Through the low-power binocular microscope, he watched them swim in the seawater of his culture dishes with a whirring border typical of surface cilia. Larvae of the maternal ascidians are not ciliated. He observed how they grew progressively until they were about a millimeter long and just visible to the naked eye. This was the stage when they should be developing into juvenile adults—but through all of his patient observation they failed to do so. "I kept them for over a month, with some of them dying off all the time, and eventually they all died without any of them metamorphosing." This endpoint to the experiment was bitterly disappointing, though, to a detached observer, Williamson's results in such few experiments over such a short space of time would appear truly astonishing. "At this time, I was preparing a more comprehensive paper on this experiment. I knew, of course, that I would have difficulty getting anybody to believe me."

On March 30, 1990, Williamson traveled to Boston at the invitation of Lynn Margulis. He had the pleasure of meeting the evolutionary biologist for the first time before presenting his theory at the Boston Colloquium for Philosophy of Science. The title of his talk was "Sequential Chimeras." He took the stage before a gathering of some 30 people in the cavernous Boston University Law School auditorium, built to house an audience of 500.

Quoting Homer, he explained his use of the term *chimera*, which derived from the Greek fable of a fire-spouting monster, with a lion's head, a she-goat's body, and a serpent's tail. The term had been adopted in more prosaic fashion to describe the experimental blending of tissues from two different species, or of two genetically distinct strains of the same species, by techniques such as grafting or genetic engineering. "According to most dictionaries," he noted, "chimera has another definition, one often regarded as the most common usage, where it refers to a mere wild fancy, an unfounded conception, an absurd creation of the imagination." He added, with a chuckle: "I should now like you to consider the possibility, indeed the strong probability, that organisms with identifiable features of different origin existed long before there were genetic engineers or even ancient Greeks." In this way he introduced them to the idea that a great many animals (and plants too, though he wasn't to address this in the lecture) are really chimeras, produced by nature over the vast periods of evolutionary time. Unlike the grafted rose or the mythical monster, these chimeras did not readily reveal their miscegenetic origins; rather, their blended ancestry manifested itself, like successive acts of magic, during consecutive stages of their life histories. He felt it not unreasonable to refer to them as "sequential chimeras" since they began with the features of one group of animals before metamorphosing to an entirely different group, frequently a different phylum. Indeed, if Williamson's reasoning were to be accepted, Haeckel's elegant evolutionary tree was about to become extinct.

His audience responded with predictable incredulity. Like Williamson himself, they had been steeped in orthodox views since biology classes at school. They were well aware that Darwin saw larval forms as important clues to understanding the evolutionary past. For Darwin, the conclusion was inescapable: "As the embryo often shows us more or less plainly the structure of the less modified and ancient progenitor of the group, we can see why ancient and extinct forms so often resemble in their adult state the embryos of existing species of the same class." In *The Origin*, Darwin had illustrated this with "what we know of the embryos of mammals, birds, fish and reptiles—that these animals are the modified descendants of some ancient progenitor, which was furnished in its adult state with branchiae [gills], a swim-bladder, four fin-like limbs, and a long tail, all fitted for an aquatic life."[3] On this point Williamson agreed with Darwin. The development of echinoderms and other phyla really did provide clues to their evolutionary development and taxonomic relationships. "But if we are to accept Darwin's viewpoint, echinoderm larvae have left-right symmetry because, at an early stage in the evolution of the entire phylum, the adults also had left-right symmetry. This leads us to believe that the larvae have retained this while some of the adults have evolved radial symmetry. Meanwhile we have to presume that others, which might have retained the left-right symmetry, have become extinct."

This was what his audience actually believed. But then he went on: "I subscribed to this myself for many years and taught it to my students, but now I have changed my mind."

He invited them to consider the fossil record of the echinoderms, which was plentiful despite its great antiquity. Many bizarre and beautiful early forms had been found, some resembling plants, complete with stems and roots, or spiral shells, or even handheld mirrors—like the most wonderful, and beautiful, experiments in three-dimensional sculptural form. But while a proportion of these had three arms and some five, none were bilaterian. To this Williamson added that no larval forms have been found in echinoderm fossil strata from the Cambrian deposits, dating to more than 500 million years ago. Though biologists might assume that this was the result of their soft bodies leaving no fossil impressions in rock strata, other soft bodies had left striking impressions. Yet it fitted perfectly with his proposal that the echinoderms had not acquired larvae until the Carboniferous, some two hundred million years later, a time when echinoderms were indisputably radially symmetrical. To put it bluntly, the conventional hypothesis for the evolutionary origin of echinoderm metamorphosis was, when you examined it closely, fraught with difficulties.

"I'm well aware," he confessed, "that from the point of view of the Darwinians, this makes me a heretic. But mine is a relatively small heresy. I agree with Darwin that organisms have evolved from other organisms and that natural selection has played an essential part in determining survival or extinction. I also agree with him that evolution is not restricted to adults: embryos and larvae too have evolved and are still evolving. I have, however, come to disagree with his view that embryos and larvae must, in all cases, have the same genealogy as the adults they give rise to."[4]

Williamson would subsequently deliver this same presentation to members of the Marine Biological Laboratory at Wood's Hole, one of America's leading marine biology research facilities. Speaking calmly and elegantly, peppering his talk with humor, he hardly cut the figure of a heretic. But in Tauber's memory, "To the experts who were present, Don's hypothesis was far-fetched, if not simply outrageous."[5]

Williamson reminded his second audience that he was not the first to adopt such heretical views. Fell, the eminent New Zealand expert on echinoderms, had arrived at the same conclusion as long ago as 1948. Observing that larval morphology was full of incongruities, Fell recommended the very opposite of Darwin—that biologists should focus on adult morphology alone, and disregard the larval evidence, when deducing how the phylum had evolved. While Fell had made no convincing attempt to explain the incongruities between adults and larvae, Williamson offered a logical and consistent synthesis. "I postulate that the first echinoderm larva originated in another phylum, that its acquisition by an echinoderm did not occur until after all the existing classes were well established, and that the larva gradually spread from one individual to another, species to species and group to group within the phylum, evolving as it spread."

Williamson went on to illustrate what he meant with the example of *Astropecten aurantciacus*, a gold-colored starfish that inhabits coastal waters, and one that was very familiar to the marine biologists.

"The adult is a five-armed star that crawls over the ocean floor. It is radially symmetrical. The larva on the other hand is bilaterally symmetrical. Why, then, should *Astropecten*, like thousands of other echinoderms, spend the early part of its life as a bilaterally symmetrical larva and then switch, dramatically and totally, to radial symmetry for the remainder of its adult life?" What, in other words, was the advantage of bilaterian symmetry to a free-floating organism? Would it not have made more sense the other way round, with the radial animal floating freely and the bilaterian animal crawling over the ocean floor? He showed his audience a picture of the *Astropectan* larva, which looked, in silhouette, like a sports trophy's two-handled cup. The larva is called a "bipinnaria" because it is equipped with two rings of locomotory cilia. "This," he explained, "is the typical larval form of the majority of starfish. According to Darwin, any other marine creature with this type of larva must share a common earlier ancestry with the starfish." He projected onto the screen the very similar tornarial larva of the acorn worm. In the acorn worm's phylum, as Williamson freely admitted, the development from larva to worm had none of the catastrophic violence done to the bipinnaria of the starfish during its metamorphosis to adult. "There is no quasi-parasitic behavior. Instead, the adult really does develop from the juvenile, much as a human fetus does from the embryo, by a process of extension and differential growth." Here were two very similar, potentially related, larvae, each participating in the metamorphic life cycle of very different phyla, going on to very different patterns of metamorphic development. In his opinion, this made no sense at all when interpreted through the orthodox perspective.

He shifted ground to pose a hypothetical question. What if there were no larval similarities between the echinoderms and hemichordates? Would two such dissimilar phyla ever have been linked to the same branch of the evolutionary tree? Of course he knew that they would never have been linked. "It is time," he concluded, "that we chopped down this phylogenetic tree, since the clues from embryos and larvae have been just as badly misinterpreted in the other main branch."

Biologists needed a new theory, he suggested, one that reconciled the similar larvae as well as the extremely dissimilar adults of echinoderms and sea acorns. He then went on to explain his theory of larval transfer, adopting the example of the echinoderms.

Originally the echinoderms had no larvae. They were radially symmetrical throughout life. At some subsequent stage the echinoderms incorporated the genetic prescription of a bilaterian animal, such as the acorn worm—an ancestor that subsequently developed to a larval stage similar to an acorn worm's tornaria. This led to the familiar echinoderm metamorphosis, which accommodated the two developmental blueprints. Where the acorn worm plan was bilaterian throughout the life cycle, the echinoderm metamorphosis inserted the prescription for a bilaterian life history into a preexisting blueprint for radial symmetry.

The gastrula stage, where the embryo had evolved the beginnings of a mouth and gut, was the key to this development, and to the configuration of the cells that would become the newly forming echinoderm. In Williamson's opinion, early echinoderms, without larvae, would have developed their pentaradial shape from the blastula, as in Kirk's brittle star. After an echinoderm and a hemichordate hybridized, the hemichordate portion of the genome took control of development before the echinoderm half, and it produced three pairs of abdominal, or "coelomic" sacs, all composed of stem cells. When one of these sacs reached the right size, the echinoderm portion of the genome recognized it as similar to a blastula, and it proceeded to make a pentaradial echinoderm from it. In this way, the first echinoderm to acquire a larva added a bilaterian planktic dispersal phase to its life history. This wider dispersion gave many echinoderms an adaptive advantage over their non-metamorphosing relatives, which had no such easy method of escape or dispersal in times of starvation or environmental stress. Natural selection would have selected for the metamorphosing species.

This concept of whole genomic incorporation, with its potential for dramatic metamorphosis between life-cycle stages, Williamson labeled his "sequential chimera" theory. He was unapologetic in affirming that his theory of the acquisition of larval genomes was an example of non-Darwinian evolution, involving a reticulate or blending pattern of evolution, as opposed to the classical Darwinian branching pattern, and it assumed a sudden, or saltationist, evolutionary event as opposed to the classical gradualist pattern. Indeed, he speculated that this evolutionary mechanism went much further. Animals in many other phyla have incongruous larvae, which suggested that they too had been later additions to the life histories in which they now occurred. Nowhere was this more apparent, or interesting, than in another familiar form of larva, known as the trochophore.

To a nonprofessional eye the trochophore somewhat resembles an acorn with a furry hatband and a Mohican-style apical tuft of hair. In fact the term *trochophore* is taken from the Greek for "wheel," since the furry hatband is, in reality, a tirelike equatorial ridge, bearing a ring of locomotory cilia, which causes the larva to spin like a wheel as it moves through the planktic waters. Trocophores are found in four different phyla, including annelid worms, spoon worms, peanut worms, and mollusks, this latter, the third largest phylum among the animals, with some 50,000 species, and including the snails, slugs, clams, cuttlefish, and the octopuses we saw earlier. Mollusks are among the most important, and varied, of the marine invertebrate animals, dating back to the earliest Cambrian strata, and perhaps even earlier still, to the first soft-bodied animals of the so-called Ediacaran period, between 570 and 600 million years ago. Williamson drew the attention of his audience to the fact that two whole classes of mollusks have no larvae, including the cephalopods, such as cuttlefish and octopuses. Meanwhile trochophores are found in some but not all of the remaining classes of mollusks, and variants of the trocophore are also found in the phylum of the nemertine worms. In fact, trochophore variants occur in at least seven different marine animal phyla, where some metamorphose to juveniles while others metamorphose to another, radically different, type of larva. Did this imply that all of these phyla shared a common ancestor? And if so, how did one explain that other mollusks develop through quite a different larva, known as a pericalymma?

Williamson went on to describe the catastrophic changes that accompanied metamorphosis in many marine animals, where independent development of the juvenile adults was reminiscent of what was seen in the echinoderms. To him, many of these observations made no sense from the perspective of linear descent with modification. But if his theory was right, it offered a rational explanation for such anomalies, with implications every bit as startling for genomic evolution as witnessed in the gross physical violence seen in the body and tissues of the animals. He did not deny that his theory invoked difficulties that still needed to be overcome—difficulties of both a theoretical and practical nature. "I admit," he concluded, "that the full implications cannot be assessed until the extent and limitations of larval transfer are defined." In particular, the genetics of such hybrids needed to be evaluated. Complex genomic rearrangements must come into play, especially so when the parental generations were widely separated in the evolutionary tree. What form these rearrangements would take depended on genomic mechanisms he could neither imagine nor predict.

Williamson's lecture fell like a bombshell upon his listeners. In Tauber's recollection: "I rose up and assessed the audience—which included some apoplectic sea urchinologists. Either we were all privy to a historical moment, or else Williamson would be recorded as an ordinarily competent and diligent zoologist, who lived with animals by the seaside for thirty years, and whose mistake should be forgiven in the light of his standard contributions." In Williamson's more phlegmatic recollection: "There were a few questions at the Boston symposium but a lot of questions at Woods Hole."

He came away with an overwhelming impression of disbelief. Several responded with outright rejection of his ideas. "They told me flatly that they just didn't believe it. It couldn't happen. I was nuts. I must have mixed up my cultures."

The referee who had turned down his submission to the *Proceedings of the National Academy of Sciences* as not worthy of a competent high schooler was in the audience. He now took to his feet and confronted Williamson:

"This is impossible."

Williamson would subsequently recall: "His name was Dick Whittaker. He was a marine biologist doing research on ascidians, and he knew a lot more about ascidians than I did."

"Very well," Williamson countered. "This is a marine station. Presumably you have sea urchins and ascidians here. Let's try the same experiment here and now. Let's start tomorrow morning."

Whittaker was surprisingly game. He duly produced some urchins and ascidians. But when they attempted a new cross-phyletic hybridization, they obtained no living offspring. Williamson felt, however, that Whittaker was finally convinced that the English biologist was not talking through his hat. One or two eggs divided once, but then they stopped. This suggested to Whittaker that it could just possibly happen. Williamson's American lecture was published as a chapter entitled "Sequential Chimeras" in a book edited by

Tauber, *Organisms and the Origins of Self.* Tauber commented in the editorial that the incredulous reception Williamson received in Wood's Hole was only to be expected. But the very essence of science was the development and testing of ideas and theories, and Williamson's theory was at the very least stimulating debate and experiment. Tauber summarized a number of questions that needed to be answered. More photography and anatomical dissection of the hybrid offspring was needed. Ascidians were hermaphrodites and might thus self-fertilize. Future experiments would need to rule this out with particular detail. Williamson might well counter that the offspring were not those you might expect of an ascidian mother, whether through hermaphroditic or sexual fertilization, but of the echinoderm father. But there was no doubt about the relevance and importance of Tauber's final suggestion. Detailed genetic confirmation was essential if the hybridization results were to be believed.

Williamson returned to the Port Erin Marine Laboratory in early May 1990, determined to resolve these questions.

— 7 —

Catastrophe

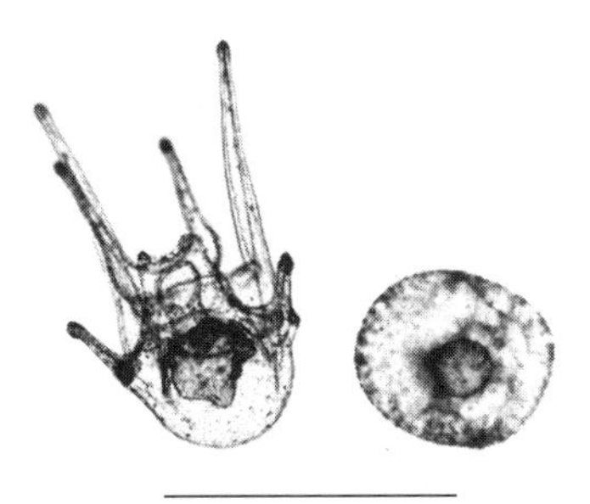

Metamorphosis of hybrid pluteus to spheroid

In February 1990, and prior to his American trip, Williamson had initiated a new cross-phyletic hybridization experiment, fertilizing another batch of eggs from the sea squirt, *Ascidia mentula*, with sperm from the sea urchin, *Echinus esculentus*. In keeping with his increasing expertise in hybridization methods, this experiment yielded an enormous number of successful hatchlings, three thousand of which would successfully metamorphose. As previously, the hatchlings developed not to the tadpole larva of the maternal ascidian but to the paternal pattern of pluteus larvae. He fed the huge harvest on a diet of diatoms and kept them alive in several glass bowls of filtered seawater, changing the seawater twice a week. By March 12, the larvae were a little more than a twenty-fifth of an inch long—relatively large for a planktic larva. The majority then underwent a curious change, resorbing their ciliated arms and the internal skeleton that supported the arms. Williamson monitored this change, as they rounded off into spherically shaped offspring that buzzed about the water table through the action of surface cilia. He called these strange new organisms "spheroids." Earlier hybrid researchers had also reported rounded shapes but these had not developed from plutei, and Williamson thought these earlier observations were probably the results of polyspermy, a developmental anomaly resulting from the fertilization of the ovum by more than one sperm. He also noticed that his spheroids developed a little bump on the surface. This appeared to be a specialized organ, a suction cup that enabled them to attach to the walls of the glass bowls, where they would remain affixed for hours at a time. He magnified the spheroids and took photographs to record their appearance.

When I first saw microscopic specimens and pictures of the spheroids in January 2002, I asked Williamson if these organisms with their ciliated surface and suction cups had ever been reported before. He shook his head. "No," he declared. "These really were new. They were neither tadpoles nor plutei—although they had metamorphosed from plutei. Such things have never been described in echinoderm development at all."

I was aware that sea squirts, unlike urchins, do attach themselves to rocks and other solid surfaces, though not, to my knowledge, through the action of suction cups. But the tube feet of the sea urchins did employ suction cups.

"Could these spheroids," I pressed, "represent the original larval stage of the sea squirts before they acquired tadpoles?"

Again, he shook his head. "I couldn't determine this one way or the other."

The majority of the hybrid larvae metamorphosed to spheroids, but in a smaller number—perhaps seventy or so—the pluteus morphology persisted and these developed "rudiments" within their coelomic pouches. In normal sea urchin plutei, these rudiments would signify the start of metamorphosis to the juvenile adult urchin. In these seventy or so individuals, juvenile adults developed into the familiar pentaradial form. Williamson observed these changes, noting that they followed the normal pattern of echinoderm metamorphosis. He was thrilled when the tiny sea urchin juveniles crawled out from the discarded wrecks of the dead plutei that had been sacrificed for these beautiful yet curiously uncaring offspring.

When Williamson had traveled to Boston, he had left this ongoing experiment in the capable hands of his colleague, Alan Bowers. The large numbers of spheroids had been kept in their bowls while the hybrid urchins had been transferred to a small Perspex aquarium tank, kept at 59°F (15°C) and continuously irrigated with a trickle of filtered seawater. After his return home, Williamson took up the experiment from where he had left it, moving the surviving spheroids into the same tank as the urchins, meanwhile keeping a few in bowls to see if they would change into anything else. The urchins continued to grow, but the spheroids retained their size and form, suggesting that this change was an end point in their metamorphosis. Williamson now counted some twenty free-living sea urchins. He continued to observe both sets of offspring, while collecting more specimens to repeat the experiment. As usual, he was limited by the urchin breeding season. Meanwhile, he began to write a new paper, intending to report the data he had amassed, while refuting the suggestion of some among his American audiences that he might have mixed up his cultures. He had, of course, taken ample precautions to avoid mixing the cultures and was confident that this could not have happened. He also planned some new experiments.

While collecting a bucketful of fresh urchins on a rocky shore, he jumped down from a boulder and slipped on the rocks, striking his head. It seemed no more than a minor accident, resulting in an inch-long gash. He drove back to the lab, where he put the urchins into a seawater tank. As he was doing so, his right arm felt as if it was misbehaving. "I had some movement, but I was not in full control of the movement. My whole arm felt awkward and clumsy. I realized something was wrong. So I headed for home. But along the way I lost the use of this arm entirely and meanwhile I was also losing control of my right leg. The car had an automatic gearbox, so I was able to use my left foot on the accelerator, eventually turning into the drive with one hand. I couldn't complete the lock and ended up stuck half in and half out. I knew by now that I was having some sort of stroke."

His wife, Enid, rushed to his assistance and called the family doctor. By the time he arrived by ambulance into the emergency room of Noble's Hospital, Don could hear and understand what people were saying to him but he couldn't voice a reply. When he did try to speak, his words emerged as gibberish.

The doctors confirmed that Don had suffered a cerebral hemorrhage. At a critical moment in his hybridization experiments, the iconoclast had lost his voice.

It was a desperate tragedy for Don Williamson, lying shocked and dispirited in his bed in the island capital of Douglas, wondering if perhaps he would ever be able to continue his extraordinary line of research. We shall return to the story of this dedicated scientist and his extraordinary quest. But for the moment let us turn from the exotic world of marine invertebrates to the equally exciting and colorful world of insects, where many distinguished thinkers had long shared the marine biologist's fascination with the strange and beautiful mystery of metamorphosis.

PART TWO

The Butterfly's Tale

You have a story to tell. Tell it to me; and for a year, for two years or longer, until I know more or less all about it. I shall leave you undisturbed, even at the cost of lamentable suffering to the pines.

But let me tell you: we are called old-fashioned, you and I, with our conception of a world ruled by an Intelligence, we are quite out of the swim. Order, balance, harmony: this is all silly nonsense. The universe is a fortuitous arrangement in the chaos of the possible. . . . Chance has decided all things. . . . It hardly seems so. The riddle is as dark as ever.

—JEAN-HENRI FABRE,
The Life of the Caterpillar

8

The Evening of the Great Peacock

Great peacock moth

It is late-nineteenth-century France, nine o'clock, a warm evening on the sixth of May. At his home, the Harmas de Sérignan, the naturalist Jean-Henri Fabre is preparing for bed. Suddenly there is a commotion in the adjacent bedroom, where his son is supposed to be sleeping. Little Paul, half undressed, can be heard rushing around his bedroom, jumping and stamping, knocking the chairs over like a mad thing.

"Come quickly, Papa!" he shrieks. "Come and see these moths, big as birds. The room is full of them."

Fabre rushes into his son's bedroom to witness an unprecedented event even in his wide experience of the strange and wonderful world of insects. Little Paul's bedroom has been invaded by swarms of gigantic moths. Four are trapped in the birdcage, and many others are fluttering their wings against the ceiling.

"Put on your things and come with me!" Fabre exclaims.

In a bustle of activity, Fabre helps his son dress, and they hurry downstairs to the study, which occupies the right wing of the house. In passing through the kitchen, they have to skip around the frightened servant, who is flapping her apron at swarms of the same invaders. Grabbing a candle on its sconce, they enter the study, where the scene resembles a wizard's cave. Swarms of the giant moths fill the air, spiraling and eddying like a swarm of bats. With a soft flick-flack they dart and knock around the bell jar Fabre had that very morning placed on his laboratory table. Inside the jar is a female moth of the same species as the invaders, a magnificent great peacock. Fabre had watched her emerge from the cocoon he had collected from the bark of an old almond tree. He had imprisoned her, still damp with the humors of hatching, under the wire-gauze bell jar, not because he was planning any specific experiment, but merely to indulge his curiosity of nature. But now it looked as if this had triggered the male invasion.

The great peacock moth, *Saturnia pyri*, has a broad distribution through the warmer regions of Mediterranean Europe, extending south to North Africa and east to the Middle East. It is spectacularly attractive, with a wingspan that can exceed six inches—almost as wide as the open adult hand. "Who does not know this magnificent moth," writes Fabre of the male of the species, "clad in maroon velvet with a necktie of white fur, its wings, with their sprinkling of grey and brown, crossed by a faint zigzag and edged with smoky white, in the center a round patch, a great eye with a black pupil and variegated iris containing successive black, white, chestnut, and purple arcs." Now, suddenly and unexpectedly, these exotic giants had taken possession of his home. And over the days that followed, the visitors continued to flock to the bell jar, in what appeared to be extraordinary numbers. Fabre would later describe his astonishment in his exquisitely perceptive book *The Life of the Caterpillar*:

> The aggregate of the visitors during those eight evenings amounts to a hundred and fifty, an astounding number when I consider how hard I had to seek during the following two years to collect the materials necessary for continuing these observations. . . . For two winters I visited every one of those decayed trees at the lower part

> of the trunk, under the tangle of hard grasses in which they are clad, and time after time I returned empty-handed. Therefore my hundred and fifty moths came from afar, from very far, within a radius of perhaps a mile and a half or more. How did they know of what was happening in my study?[1]

With this question began one of Fabre's now-famous experiments, investigating how the male moth tracks down his mate from great distances, negotiating stormy weather and every obstruction. Vision could hardly be the answer. The sexual seduction, and that is exactly what he was observing, took place over too great a distance and in the darkest night. "No—the path drawn by the male in search of the female is far too erratic for any visual signal, leaving only the possibilities of scent or sound. . . ."

Today, more than a century after Fabre made his original observations, his descriptions remain the most enchanting ever made in the world of entomology—and some believe, in the entire realm of nature. Readers wishing to discover more can explore his book, in which they will discover that the sole purpose of this sumptuous grace and beauty—the three days of life and obsessive purpose of the adult male moth—is a ceaseless search for the female and the faintest hope of mating.

In *The Descent of Man*, first published in 1871, Charles Darwin, a Fabre contemporary, pioneered the concept of sexual selection as an evolutionary force.[2] Mammalian females brood their young within their bodies. Even after birth, whether the offspring first enters the world as a still gestating egg or infant, the female is often the parent with the heaviest responsibilities. Females have a vested interest in the genetic health and vigor of their offspring. Evolution is all about passing on the genes of inheritance, and nowhere is this more focused than in the act of mating, with its potential for offspring carrying those genes. It is not surprising, then, that females are rather choosy about their mates. All those wonderful rituals of courtship arise from this: the jousting of stags, the tail of the peacock, the red breast of the perky robin in our garden, and the song of the lyrebird. Each male endowment is the result of female sexual selection pressure over vast periods of time. Yet Nature, in its biological strategies, can be altogether ruthless. Some female praying mantises, including a British population of *Mantis religiosa*, will cannibalize their partners, headfirst, during copulation.[3] This same ruthlessness has condemned the adult male of the great peacock moth to his uncontrollable desire to find a mate, his life encapsulated to a few days of frantic searching. His mouthparts are so rudimentary, he cannot eat or drink. The bulk of his energy, inherited from the dedicated feeding of the caterpillar stage, has been diverted to the production of his vainglorious body, bedecked with the four giant wings, each carrying the distinctive peacock eye.

In his investigation of the male's courtship flight, Fabre paid particular attention to the quadruple antennae, massively enlarged when compared with those of the female. She, on the other hand, "appears equipped for seduction with a song as irresistibly enchanting as that of the sirens in the voyages of Ulysses."

But was it really a siren song?

In fact a song of any sort was soon ruled out of the question. "The great fat [female] moth, capable of sending a summons to such a distance, is mute even to the most acute hearing," as Fabre now realized. His first big clue came from the fact that he could halt the attraction of the males toward the females by cutting off the male moth's antennae. This suggested that scent might be the answer. "Are there, in point of fact, effluvia similar to what we call odor, effluvia of extreme subtlety, absolutely imperceptible to ourselves and yet capable of impressing a sense of smell better-endowed than ours?"

Fabre was more than just a poet of nature: he was a master of scientific experiment. And in such "effluvia of extreme subtlety" he had indeed discovered the solution to the mystery. Today we know that the female great peacock moth attracts the males from a mile or more distance through exuding the equivalent of an extremely subtle scent—her siren call is a chemical of great potency known as a pheromone, released by scent-forming patches on her abdomen. The male is exquisitely sensitive to her pheromone, which he detects, exactly as Fabre discovered, by using his richly developed antennae.

The pheromones of insects have practical importance. In silkworm moths, for example, where the active substance has been extracted from the tail ends of the abdomens of some half a million females to be isolated in a pure enough state for chemical analysis, it was found to be a simple alcohol, with a sixteen-carbon chain. At the slightest contact with this substance, the male violently vibrates his wings. Further experiment has revealed the exquisite sensitivity of the male silkworm moth to this siren scent, when the most minuscule secretion that modern science can detect—defined as a "unit of attraction"—is in the order of 10^{-10} of a microgram. This is one-ten-thousandth-millionth-millionth part of a gram. The pheromone is picked up by the antennae of the male, and it guides him with uncanny accuracy through dark and storm to find the seductive source. When males of the Chinese saturniid moth *Arctias selene* were released at a distance of eight miles from caged females, more than one in four of them found their way to the cage. It made no difference, as Fabre himself established, if scientists placed all kinds of additional strong odors as obstacles in their path. The male was not deflected in the least. Today we know that it is not only the female that produces these subtle signals of seduction. In many male moths and butterflies specific scents are liberated by special glands that have an equivalent aphrodisiac function, in turn exciting the female to accept the attentions of the courting male.

Fabre conducted his studies not only on the bench and in the test tubes of his home laboratory but also in his *harmas*, an enclosed piece of land behind his house. In this natural sanctuary, he planted suitable shrubs and trees, otherwise allowing nature to run wild. During the summer following the evening of the great peacock he also paid street urchins to collect caterpillars of the great peacock for him, at a sou apiece. From these he reared his new menagerie on almond-tree branches in his *harmas*, watching them mature and then pupate in the cracks and crevices of the bark. That winter, assisted by his friends, he searched through the brambles at the base of the almond trees—the leaves of which were the chosen food of the caterpillars—to become the possessor of an assortment of cocoons, the bulkier and heavier of which denoted metamorphosing females, which he then would observe for a new generation.

Fabre went on to dissect some of the brown cocoons, using his pocket knife to cut open the leathery case to see what was happening inside during that long period of exclusion from the world. How, he wondered, from this intermediate stage of dormancy, did that extraordinary metamorphosis take place, leading to the emergence of the magnificent adult creature, perfectly formed from the moment of its birth, with its multifaceted eyes, its plumes of antennae, and its great wings unfurling and stretching in the warmth of the sun? Did the wingless yellow caterpillar, with its palisade of black hairs, and its beads of turquoise blue, invoke some natural magic or spell to grow those enormous wings, or those huge fleshy antennae, or the compound eyes? What he discovered was a mystery far more beguiling than might be evoked by spell or magic.

The development of the caterpillar really does take us into a realm of the bizarre and wonderful. Indeed, the apparent quiescence of the pupa could not be more misleading. Inside its leathery wall, the living being of the caterpillar does not change by the addition of wings or eyes to its former body. Instead the body of the caterpillar melts to an organic soup of cells to begin its biological construction all over again.

For the deeply religious Fabre, it must have appeared for all the world as if he were witnessing a miracle of nature in which the newborn insect, with its multifaceted eyes, articulated legs, newfound sexual maturity, and extraordinary wings, emerges from the sacrifice of its former self in what amounts to the real-life enactment of the mythical phoenix arising from its own funeral ashes.

How could such a riddle ever have come to be?

The answer would not come from the gentle Fabre, whose writings were so imbued with an enchantment of description he has been labeled "the Homer of Insects," but from another genius with an equally profound interest in nature, that English pioneer of evolutionary biology, Charles Darwin. Ironically, Fabre would refuse to contribute to the evolutionary aspects of its ultimate solution, despite being invited to do so by Darwin himself—and the reason for his refusal lay in the very poetic nature of his genius.

Darwin's landmark book on evolution, *On the Origin of Species by means of Natural Selection*, was published in 1859—a generation prior to Jean-Henri Fabre's original French publication of "The Great Peacock" in *Souvenirs Entomologiques*. We know that Fabre read *The Origin* just as we know that Darwin read Fabre's *Souvenirs*. Darwin made no secret of his admiration for Fabre, not only for the beauty of his descriptions but also for the precision of his experimental method. But Fabre did not reciprocate with regard to Darwin's evolutionary theory, a rejection that must have dismayed the aging English naturalist.

In a letter dated January 31, 1880, Darwin wrote to Fabre to suggest, "I am sorry that you are so strongly opposed to the descent theory; I have found the searching for the history of each structure or instinct an excellent aid to observation; and wonderful observer as you are, it would suggest new points to you. If I were to write on the evolution of instincts, I could make good use of some of the facts you give."[4]

But even this respectful appeal failed to persuade Fabre, who was irreconcilably opposed to the very concept of evolution.

In chapter 8 of his book *More Hunting Wasps*, first published in English in 1919, Fabre described a "nasty and seemingly futile" experiment in which he reared caterpillar-eating wasps on a "skewerful of spiders." To this he added the explanation: "I should not have undertaken these investigations, still less should I haven spoken about them, not without some satisfaction, if I had not discerned in the results . . . a certain philosophic import, involving, so it seemed to me, the evolutionary theory."

Whereas the diet of the caterpillars was omnivorous, Fabre

noted, the parent wasp was limited to a single kind of prey, whether a particular kind of cricket, or locust. Assuming that the diet of the ancestor of the hunting wasps was omnivorous, like the caterpillar, he demanded to know why, through evolution—"that pitiless fight for existence that eliminates the weak and incapable and allows none but the strong and industrious to survive"—the descendant wasps had given up their eclectic advantage to depend on such a limited diet, which must often threaten starvation. In Fabre's own words, "It is assuredly a majestic enterprise, commensurate with man's immense ambitions, to seek to pour the universe into the mould of a formula. . . . But . . . in short, I prefer to believe that the theory of evolution is powerless to explain [the wasp's] diet."

From the specific example, Fabre extrapolated a generalization. Darwin's evolutionary theory was a mathematical contrivance through which "one was merely observing the same idea from different points of view." He assumed, incorrectly, that the mathematical reductionism of evolutionary theory demanded an ideal situation, which never prevailed in nature. "Yes, it would be a fine thing to put the world into an equation. . . . Alas, how greatly must we abate our pretensions. The reality is beyond our reach when it is only a matter of following a grain of sand in its fall; and [despite all this] we would undertake the ascent of the river of life to its source!"

Yet in such profound disagreement between the two great naturalists there is also something touching and poignant. Fabre could hardly be dismissed as ignorant of biological science. What he represented was a different philosophical tradition harking back to an earlier age, a more innocent age, perhaps, with respect to its views of the divine nature of creation and the balances of nature. But their disagreement did not translate to personal animosity or bitterness. In the words of Fabre's friend and biographer, Dr. G. V. Legros:

> It seems that on his side Fabre took a singular interest in the discussion [exchange of letters with Darwin] on account of the absolute sincerity, the obvious desire to arrive at the truth, and also the ardent interest in his own

> studies, of which Darwin's letters were full. He conceived a veritable affection for Darwin, and commenced to learn English, the better to understand him and to reply more precisely; and a discussion on such a subject between these two great minds, who were, apparently, adversaries, but who had conceived an infinite respect for one another, promised to be prodigiously interesting.
>
> Unhappily death was soon to put an end to it, and when the solitary of Down [Darwin] expired in 1882 the hermit of Sérignan saluted his great shade with real emotion. How many times have I heard him render homage to this illustrious memory![5]

One can only sympathize with both men: with Darwin in thinking that Fabre had the genius and perception to open the wonder of insect metamorphosis to evolutionary understanding; and with Fabre, whose Homeric vision and religious awe at the complexity and beauty of nature rebelled against the very concept of evolution. Alas, posterity has confirmed that Fabre was as mistaken about Darwin as Keats was about Newton. Fabre was wrong in his interpretation of how natural selection works—natural selection in adapting an adult insect to a single source of food does not pretend to lead by any route to perfection, only to signify adaptation to a specific life cycle, or ecology, or sexual selection by the partner. And science, whether investigative or evolutionary, does not erode our wonder at the beauty of the natural world. On the contrary, as Fabre himself amply demonstrated, it enhances it.

In fact, Darwin's vision threw open a new window onto the understanding of life, the wonders of its origins, its beauty and diversity. And while the gentle Fabre refused to contribute his understanding of insects to this vision, there would be others, distinguished in their own right, prepared to take on that challenge. Foremost among these would be another English naturalist.

Vincent Brian Wigglesworth was born into a medical family in Kirkham, a small town near Preston in Lancashire, on April 17,

1899, seventeen years after the death of Darwin. In Provence at this time, Fabre was a youthful seventy-six, writing the books that would make him famous far beyond the field of entomology and never ceasing to "multiply his pinpricks in the vast and luminous balloon of transformationism [evolution] in order to empty it and expose it in all of its inanity." Both Fabre and Darwin would inspire the youthful Wigglesworth's love of insects, an inspiration that, however tortuous the road, and spanning almost the entirety of the twentieth century, would lead him to solve some of the mysteries of the butterfly's tale.

9

The Science of Life

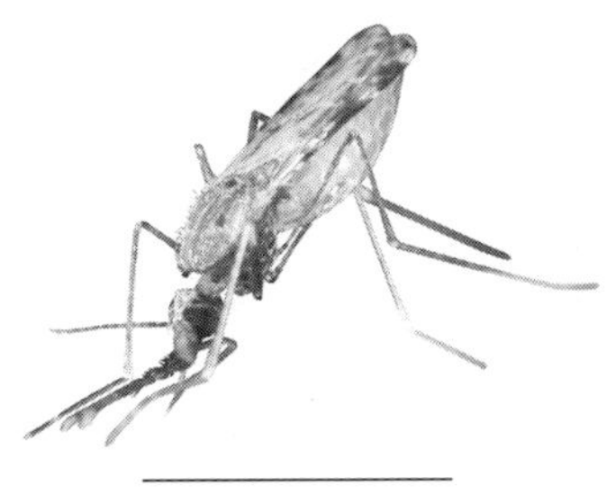

Malaria mosquito

Anyone who has been privileged to observe young children in a garden cannot fail to notice their attraction to the miniature world of marching ants, buzzing bees, creeping caterpillars, and the colorful flight of dragonflies, moths, and butterflies. The youthful Vincent Wigglesworth was no exception. "By the age of five," as he later recalled, "I was keeping a large collection of caterpillars and other insects and spending hours and hours watching them."[1]

"In 1905, while caring for my cultures of *Abraxas grossulariata*, I discovered the metamorphosis of insects. . . . A caterpillar I had imprisoned in a jam jar wrapped itself in silk and then, after a few days, emerged under my close and astonished observation, a butterfly!" In fact it wasn't a butterfly, as the boy imagined, but a beautiful moth, variously known in the British Isles as the magpie, currant, or gooseberry moth. The caterpillars are a conspicuous pale green with bold black spots and a rusty line down the sides—a flarelike warning that they are likely to be distasteful to predators. Feeding on blackthorn, hawthorn, currant, and gooseberry bushes, the moth overwinters in caterpillar form. In England it pupates in

May or June and metamorphoses, as the youthful Wigglesworth had observed, in July and August to the striking adult moth with piebald wings of black and grayish blue, with a yellowy orange stripe in the middle of its forewings and this same flare of color spreading, mantlelike, over the back of its head and shoulders.

It is hardly surprising that the emergence of such beauty from a chrysalis was such a wonder to the then seven-year-old, who assumed he had discovered his very own fabulous secret. "Of course I found in later years that this had already been known by Aristotle. . . . Natural philosophers and small boys apparently find the same amazement in the transformation of insects from caterpillar to chrysalis to butterfly."[2]

In the family garden was a willow tree whose leaves were invaded by red swellings, known as galls. He cut open the galls and was astonished to find that each gall contained a grub. It was another mystery he was determined to solve. But as he probed the galls he found no hole, or opening, through which the grub could have crawled into its minuscule home within the leaf. "I sought the guidance of my parents about this mystery—but in vain. From that moment I was a committed entomologist."[3]

In 1911, the family moved from Lancashire to the south of England. "For me it was like moving to El Dorado—so rich in insect life were the lanes and woods of Hertfordshire in those days."[4] Wigglesworth's fascination continued to his boarding school, Repton, where he devoted his leisure time to pursuing insects in the countryside. By now it seems altogether likely that Fabre was his inspiration, for he subsequently remarked: "Although we had teaching in physics and chemistry, we had none in biology. In fact I would have felt it obscene to have these matters, about which I felt so deeply, treated as a school subject." This, for a typical English gentleman, who would subsequently be described as "gentle, reserved and formal, and with a wry sense of humor," is as close as we are going to get to his revealing a reverence as profound as Fabre's for his subject.[5]

In those early decades of the twentieth century, there were many patches of *harmas*-like wilderness to be found in the countryside around Repton where a boy might explore the world of nature. He was entranced by the immense variety of insects, which inhabit every nook and cranny of the biosphere, crawling over surfaces, burrowing through soil and decaying wood, flying through the air with a dexterity to match the most elegant birds, and even returning to recolonize the original womb of animal life, to the world of freshwater and the oceans.

Insects may be only a class of arthropods on the tree of life, but it is a heavily populated class indeed: at least five-sixths of all animal species are insects. An estimate of the total number of insects on earth runs to 10^{18}.[6] This is a million times a million times a million—too astronomical a number for the human brain to register other than in a conventional mathematical display, when it is 1,000,000,000,000,000,000. It certainly puts our own human population total of 6.7×10^9 into perspective, though we constitute a single species. At a conservative estimate the insects that make up this single class may amount to some three million species.

During his time at Repton, Wigglesworth gave some thought to his future career. His father was a family doctor, and he considered following the parental example. However, any such plans were interrupted by World War One, when week by week at the school general assembly, the headmaster would read aloud the names of pupils who had volunteered for action and were now counted in the ranks of the dead. And so, in due course, it came to Wigglesworth's turn, when he abandoned school, with just his secondary education completed, to become a signaler in the Royal Field Artillery. Fortunately, perhaps, he arrived late into the theater of war and saw little real action. Yet his love of entomology enabled him to take comfort in beauty even amid the carnage of the battlefields, as revealed by his subsequent observation that, in 1918 "I was able to watch *Papilio machaon* on the slopes of Vimy Ridge."[7] One of the creatures that also delighted Fabre as a child, *Papilio machaon* is the common yellow swallowtail, an attractive butterfly, banded black and maroon on bright yellow wings, with vivid red eyespots close to the elongated tail that gives the species its name.

After the war, Wigglesworth began his university education with his preclinical studies at Caius (pronounced *keys*) College, Cambridge, where he completed a first-class tripos in biochemistry and physiology. By now he had affirmed his decision to forgo clinical medicine—"I felt I had already seen far too much of that in the home"—and set his mind on a future in medical research. His first posting was "two very profitable years" working under the biochemist Gowland Hopkins, under whose direction he published eight scientific papers, many of these as research assistant to a recent arrival from Oxford, the eminent J. B. S. Haldane, which often involved Haldane as the experimental guinea pig in testing the effects of chemicals on himself. While conducting his biochemical researches, Wigglesworth managed to find time for a diversion into his beloved entomology, studying and publishing papers on the pigments that colored the wings of the cabbage white butterfly. In 1924, now equipped with his MD—in Britain this is a research degree and not the essential medical qualification—he moved to St Thomas's Hospital in London, where he completed his actual medical degree, the MB. Two years after Wigglesworth's move, Patrick Buxton was appointed head of the department of medical entomology at the newly established London School of Hygiene and Tropical Medicine.

Buxton was convinced that advances in infectious tropical diseases were being hampered by a lack of knowledge of the physiology of the insect carriers. He needed an entomologically trained lecturer to work with him and chose Wigglesworth on a provisional basis. Wigglesworth went straight to work. "My first discovery, in an incubator in the laboratory, was the blood-sucking bug, *Rhodnius prolixus*, which had been brought to London from Venezuela by Professor E. Brumpt a year or two earlier."[8] For Wigglesworth, this would prove a fateful happenstance. *Rhodnius prolixus* would be the subject on which he would conduct some of the most amazing experiments in the history of entomology.

But for the moment there were more pressing concerns for the young doctor-cum-entomologist. What he lacked, in practical terms, was experience of the tropics. Buxton suggested he travel to West Africa. So in 1927, Wigglesworth went to Nigeria, where he spent six months traveling before moving on to the Gold Coast and Sierra Leone.

He soon had firsthand experience of African sleeping sickness, an epidemic infection caused by a single-celled organism known as a trypansosome, which is spread by the bite of the tsetse fly. Untreated African trypanosomiasis is frequently fatal. Malaria, the greatest of all the tropical killers, was also endemic throughout the country, and bubonic plague was threatening to break out in the slums of Lagos.

Wigglesworth wrote regular letters home to a young lady, Catherine Semple, an aspiring artist whose vivacious and outgoing personality couldn't have been more opposite to his own. It would appear that she had some reservations about the serious young scientist, wondering no doubt if life might be more fun with a less academically inclined suitor. But her father, Sir David Semple, himself a pathologist, would, in due course, instruct his daughter: "Well, you know Wigglesworth is going to ask you to marry him. That young man is going places. If you don't accept him, I'm never going to speak to you again!"

In one letter home to his sweetheart, the youthful Vincent described how he helped the local American-based team: "I spent the morning with the Rockefeller yellow fever commission, while they performed autopsies on the front end of monkeys dead of yellow fever, meanwhile I collected lice from the hind end." Yellow fever, a disease in which the yellow color of the skin is caused by jaundice through massive liver destruction, is even more deadly than trypanosomiasis. It is transmitted by the bite of the mosquito and is one of the most lethal of the viral hemorrhagic fevers. In 1927 the viral cause had only just been discovered, and there was neither a cure nor a vaccine. When Wigglesworth arrived in Accra, in the Gold Coast, the main center of yellow fever research, the local situation was tense. All infected monkeys had been destroyed, and the laboratories were being decontaminated with cyanide vapors after two of the workers, the Rockefeller-based Japanese American Hideyo Noguchi and the Irish microbiologist Adrian Stokes—the latter a joint discoverer of the viral cause of yellow fever—had only recently died from the fever they were researching. They had contracted the disease from a postmortem on an infected monkey. Stokes's death in particular was a severe blow to tropical medicine, since this brave field-worker had first discovered that rhesus monkeys were susceptible to infection and, using that knowledge, had taken the first steps toward the preparation of the life-protecting vaccine that is now obligatory for travelers to the parts of Africa where the disease is endemic.[9]

Wigglesworth appeared to enjoy the experience of working in this dangerous maw, sucking up mosquito larvae out of crab holes and traveling through plague zones, accompanied by bearers carrying head loads of equipment, which included a self-contained laboratory complete with microscopes, not to mention his bed, bath, and furniture. During his travels, he discovered a number of unknown mosquito larvae, and he conducted original research on the physiology, digestion, and nutrition of the tsetse fly, which subsequently became the subjects of his first papers in medical entomology.

After his return to England, Wigglesworth was to spend many productive years as a lecturer and researcher at the London School of Tropical Medicine and Hygiene, making an important contribution to our understanding of tropical diseases and their insect carriers. He took the opportunity to broaden his travels, studying insect-borne diseases in India, Burma, Malaysia, Java, and Sri Lanka, amassing a growing reputation for entomological research. In the midst of World War Two, Wigglesworth was invited by W. C. Topley, then secretary to the Agricultural Research Council, to help search for effective insecticides.

Wigglesworth was able to negotiate terms that would enable him to work on a broad and interesting canvas. In his own words: "we should be entitled to work on all aspects of insect physiology which might reasonably be considered to have a bearing on insect control, and not only with insecticides."[10] In effect, he would be permitted to conduct "pure" research, since, as Wigglesworth saw it, all such research, in illuminating the life of insects, might in the fullness of time prove useful to agriculture—and also to medicine.

A year earlier, in 1939, the same year in which he was elected to the Royal Society, Wigglesworth published a landmark book, *Principles of Insect Physiology*, which would go through eight subsequent editions. The preface began with a statement that, in its universality of vision, was enlightening: "Insects provide an ideal medium in which to study all the problems of physiology"—the science that looks at how living bodies work. No subject is more fundamental to life, bringing together biochemistry, molecular biology, heredity, reproduction, development, and even the wonder of evolution. For Wigglesworth, the way in which insects developed, from egg to adult, including the mystery of metamorphosis, was primarily a question of physiology. His book, with its innovative breadth of thinking, would, in time, establish him as the "father of modern insect physiology."

At war's end, Wigglesworth was offered a position as head of a subdepartment of entomology within the Department of Zoology at the University of Cambridge. He accepted the position and took the Agricultural Unit to Cambridge with him. His first preoccupation was ecological: he realized that DDT, then considered a wonder invention, might, in time, fail through the acquisition of insect resistance—an evolutionary phenomenon. He was also concerned that the intrusion of DDT into complex interactions between pests and their natural enemies might disturb the balances of nature. In the early postwar years, he was reunited, in a social fellowship, with his old college, Caius, and in 1952, he was elected to the Quick Professorship of Biology, a critical step to the freedom of research that had been his dream since boyhood—much as the freedom of the *harmas* had inspired Fabre.

In attempting to explain the importance of this professional liberation, I cannot improve on the words of his protégé, John S.

Edwards, who, many years later, described his old professor as having a unique quality. "It has been said of Alexander von Humboldt that he was the last person to know all of science. . . . Wigglesworth was probably the last person to know all of insect physiology." So it was, in Edward's words, "Wigglesworth who brought together his own brilliant experiments with the work of others, from the 1930s to the 1950s, to present the first answer to the 'how' question of post-embryonic development in insects."[11]

Post-embryonic development is scientific jargon for the mystery of metamorphosis.

— 10 —

Elementary Questions and Deductions

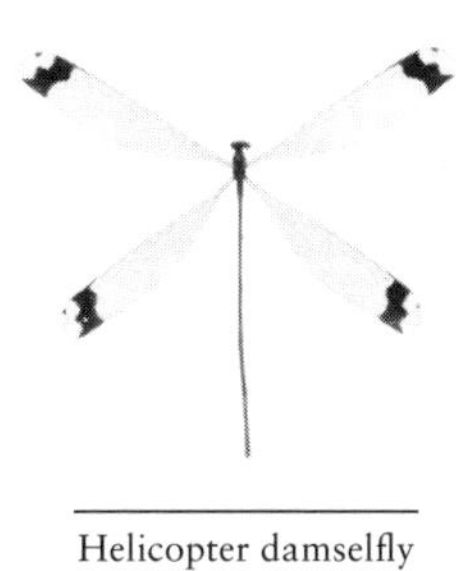

Helicopter damselfly

Wigglesworth's first scientific paper had been published in 1923, working with Hopkins and Haldane at Oxford, a study of the effects of insulin on blood phosphate. Over the ensuing ten years, now working at the London School of Hygiene and Tropical Medicine, a succession of papers reveal the progress of his interests, from purely medical to medical entomology—digestion in the tsetse fly, delayed metamorphosis in a predaceous insect, the function of the anal gills in mosquito larva, a theory of insect respiration. Then, abruptly, in 1933, the first of a classical series of papers heralded a new and all-consuming focus: an in-depth scrutiny of the mystery of insect metamorphosis.[1]

A scientist renowned for brevity in his communications, Wigglesworth often condensed his publications to less than a page. But this opening paper, addressing the theme of metamorphosis, was some fifty pages long.

It was clear from the paper that Wigglesworth knew he was confronting a very great challenge. But before he could even begin to investigate metamorphosis, he needed to take a hard look at what the mystery actually embraced. Rather like the famous Sherlock Holmes, whose author, Sir Arthur Conan Doyle, had died just three years before the paper was published, Wigglesworth probed a seemingly intractable mystery using a series of elementary questions and deductions.

His first question was obvious: why was it that all insects do not undergo metamorphosis?

What fossil evidence there was proved both surprising and pertinent: most authorities had concluded that the earliest insects did not undergo metamorphosis. The most primitive insects are found as fossils in rocks from the early Devonian period, around 400 million years ago, but none of these appears to be winged. The oldest evidence of winged insects appears to date to about eighty million years later, and these did not yet include the fully metamorphosing insects, such as butterflies, moths, bees, and beetles. This was a powerful clue, for it implied that the more dramatic changes of insect metamorphosis had entered the insect life cycle long after the more primitive adult forms had already been defined. But it made his objective more elusive. Marine arthropods, judging from their fossils, pre-date the land insects by more than a hundred million years, and we know that, at some stage, these too had evolved a diverse array of metamorphic life cycles. But now it would appear that the metamorphic life cycle of the insects had not arrived, dripping with more than a hundred million years of marine ancestry, when those ancestral forms first emerged onto dry land. Wigglesworth had no option but to go back to first principles and consider the evolutionary history of the insects. When did the ancestral species crawl out of the oceans, and what do we actually know about their subsequent history?

What a strange and alien landscape would have greeted these earliest arthropod explorers of dry land, four hundred million years ago! There was none of the familiar vegetation that clothes the modern world in its dazzling variety of grasses, forests, and flowers. No birds flitted among the trees, no dinosaurs lumbered across the plains—it was the seas that teemed with complex life: on land there were no vertebrate animals of any description. Even the continents that we recognize today did not exist. The entire landmass was joined up in a single supercontinent, Pangaea, with the green of the primitive ground-hugging plants, such as algae, liverworts, club mosses, and the stubby and knobbly Cooksonia, struggling to gain a foothold around its inclement shores. It was into this harsh landscape that those ancestors of the insects first crawled out of the oceans, struggling merely to survive before further establishing themselves in damp shady places.

Wigglesworth focused on the arthropods, assuming from what was known of early marine arthropods that the hypothetical common ancestor of this gigantic and highly diverse group of invertebrate animals would have been a marine-based creepy-crawly creature, with a body made up of ringlike segments, each supported by a pair of appendages that, for the most part, functioned as legs. The creepy-crawly's front segments fused to become the head, and the head appendages became the mouthparts and antennae. The nerves within the head developed into a ganglion, or primitive brain, which collected the information arriving from their evolving eyes, antennae, and other senses, and translated this into the appropriate movements and responses—attack, avoidance, devouring prey. Certain of the appendages at the tail end became modified for mating—Wigglesworth had looked into some aspects of this in his experiments during the African trip—meanwhile others served to manipulate the eggs as part of reproduction, or to carry sensory organs for various other purposes.

The bodies of insects were too flimsy to leave much in the way of fossil evidence. But biologists thought that the hypothetical insect ancestor would most likely have resembled a centipede, a few millimeters long. Inhabiting the shorelines, many of those early arthropods continued to evolve in oceans while others invaded the land, giving rise to the fantastic variety of creatures we see today, such as the crabs, lobsters, and shrimps we find in the oceans and the spiders, scorpions, and daddy longlegs we find on land. The insects thrived among a subdivision of the arthropods, the phylum known as the mandibulates, which, as the name suggests, use side-mounted jaws to chew their food. So successful was their colonization of land and air that the mandibulates account for the largest number of species of any animal group on the tree of life, and they still include forms that closely resemble those non-metamorphosing ancestors, such as the familiar millipedes and centipedes.

In his pioneering paper, Wigglesworth had already arrived at an elementary deduction, one that derived from this common arthropod history, and a deduction that might provide him with a first clue to the mystery: "[The process of molting] is the homologue of metamorphosis, and there can be small prospect of understanding the complex process of metamorphosis until the physiology of [molting] has been adequately described." Even those insects that did not metamorphose went through the process of molting.

Taking a careful look at this common characteristic of the arthropods, it surprised him that previous investigators had devoted so little attention to the changes that enabled molting in insects, whether or not they underwent metamorphosis. As an animal that needed to grow, much like other animals, the insect was faced with an unusual dilemma. It did not have an internal skeleton, as did a mouse or a human. Its skeleton was on the outside—thus known as an exoskeleton—and was composed of an extremely rigid structure, the cuticle, which was produced by the insect skin or epidermis. The problem facing the insect was simple: how to grow, safely and effectively, despite the restrictions of the rigid exoskeleton. Wigglesworth went on to look at what was known of this remarkable organ, the insect epidermis, and its role in the evolution and physiology of insect life.[2]

It was obvious that, in pioneering their way onto dry land, the ancestors of the insects would have needed both to conserve water and to protect their vulnerable bodies with defensive armor. They had evolved the ability to convert the carbohydrate found in mucoproteins (like the slime of slugs) into a tough fibrous secretion called chitin. Chitin is a polymer, rather like the cellulose in blotting paper. It comprises long chains of chemicals linked together to form a cohesive layer, which forms the basis of the surface covering of all insects. Over most of the insect body the chitin is toughened to make the stiff horny exoskeleton by the addition of a protein, known as sclerotin, which, when it becomes tanned in sunlight, makes it even tougher, like the shiny black carapaces of beetles. Sclerotin is also the ideal material for making the thin membranes of insect wings, or the needle-like piercing stylets of mosquitoes and the stings of wasps and bees—even the elaborate mating organs of the two sexes. The durability of this substance is made possible by chemical cross-linking sulfur groups, rather like, but much more resistant than, plastic. A similar cross-linkage of sulfur groups is the key to reptile scales, the feathers of birds, and the hair, nails, hoofs, and horns of mammals. The carapace of lobsters is hardened in a similar way, by the addition of lime to the cuticle, but even in the lobster the harder parts, the tips of the mandibles and claws, are also coated with sclerotin.

Anybody who has tried to kill a cockroach will know how tough and protective this surface armor is. It is also resistant to many harmful chemicals, even to natural decay, as archaeologists realize when they use the identity of specific insects in soil, or strata, to deduce the ecology of dig sites thousands, or even tens of thousands of years old. Moreover this surface armor has an additional, very useful property. Kept supple, for example, in the region of joints, it allows the flexion necessary for movement. In Wigglesworth's words: "Just as the invention of keratin made possible the flying equipment of the pterodactyls, birds and bats, and the hairy hugginess of mammals, sclerotin made possible the extraordinary evolutionary diversity of the insects. . . . Indeed, it is impossible to study the natural history of insects without discovering at every point how dependent they are upon this remarkable plastic."[3]

We mammals also enjoy great flexibility in our bodies and limbs, but we do so through a very different technology—an internal skeleton, equipped by a wide variety of joints, all surrounded by stretchable soft flesh. But this internalization of the skeleton also makes us vulnerable to wounding of our skin and soft tissues. The larval forms of butterflies are equally vulnerable, thanks to a soft and highly folded skin, which is necessary to allow the caterpillar to grow at incredible speed. For example, the belly of the queen termite is so elastic that, as her reproductive organs fill up with eggs, it can expand fiftyfold. The adult insect, on the other hand, dons its hard external skeleton to face the hostile world, clothing itself in this suit of armor, though with modifications in places to support the joints and the powerful jaws.

While the insect is better defended than a mammal, that same armor has one overwhelming disadvantage. A rigid outer skeleton makes further growth impossible. Unlike vertebrates, such as fish or mammals, whose internal skeleton can readily grow during childhood to attain the full adult size, insects—and arthropods in general—can only grow if they first discard their outer shells. This is the reason why crabs and lobsters—and insects—molt.

It is important to distinguish molting from the more complex concept of metamorphosis: molting merely permits an increase in size. Metamorphosis implies a change in form. The more primitive orders of insects still keep to the primitive blueprint: they molt but they do not metamorphose. These include the silverfish, which flitter across the floor from under their cozy rugs at night, the springtails, which use a forked appendage on their abdomens to leap through the air, and a number of less familiar forms that live in the humid mats under vegetation, or the moist litter of forests, or skulk under logs or stones, in environments similar to the primal ecology of those terrestrial pioneers. These simple orders can even be eyeless, and one group is devoid of the antennae so characteristic of all other insects. It was from careful observation of these primitive forms that Wigglesworth was able to make another elementary deduction.

All primitive insects are wingless. This, taken with other considerations, suggested that the mystery of insect metamorphosis might be linked to the invention of flight.

If there were seven wonders of the biological world, the flight of insects would be one of them. Ornithologists and entomologists might debate, grace for grace, the flight of a swallow with that of a butterfly, yet insects, with their extraordinary range of flight strategies and dexterities—the ability to take off backward, to hover and soar, or to alight upside down on a ceiling—is arguably more complex and inventive than that of the birds. The imagined flight of fairies, like Tinkerbell in the Peter Pan stories, more closely resembles that of an insect than any bird, much as Richard Attenborough, in his *Life in the Undergrowth* series, so charmingly demonstrated with the helicopter damselfly, the largest in the world, with wings of sapphire that extend to eight inches and a balletic grace in motion. Yet such is its heart and purpose that, in the courtship of the cascade damselfly—a rare species limited to a small number of Central American waterfalls—the male, which weighs no more than a sprite, will endeavor to impress his beloved by flying through the falling torrent.

The evolution of flight is surely one of the main reasons for the extraordinary success of the insects, helping their hunt for food, their escape from danger, their search for a mate, and their exploration, far and wide, of new habitats to make their own. In 2003 Michael Engel, an evolutionary biologist from Kansas University in Lawrence, Kansas, was researching some background information for a book on insects. Needing to describe a fossil arthropod, he peered into the contents of a dusty drawer in the bowels of the Natural History Museum, in London, and gasped at what he discovered. Fossilized in chert rock that had been dug up near Aberdeen in the 1920s, he was gazing at the remains of an extinct insect, *Rhyniognatha hirsti*, which must have been one of the earliest insects ever found. Though the fossil did not include wings, Engel and his colleague, David Grimaldi, who jointly wrote up the discovery in the journal *Nature*, found that *Rhyniognatha*, which was about five millimeters long, had jaw articulations commonly associated with modern-day winged insects.[4] Thus the fossil might represent a stage in insect evolution prior to wings, or it might represent the first step in metamorphosis of the insects. This fossil was estimated to date from 390 to 408 million years ago—an age that, if winged *Rhyniognatha* were to be confirmed in other examples of the fossil record, would push back the evolution of true insects, suggesting that the evolution of metamorphosis dated to more than 400 million years ago.

Dragonflies do not go through the stages of larva, pupa, and adult insect. But they do undergo a less dramatic change, particularly in the final molt. This is known as incomplete metamorphosis. Dragonflies and damselflies belong to the order of insects known as Odonata, which includes some five thousand species, including the giant tropical variety *Megaloprepus*, commonly known as the helicopter damselfly. This measures up to seven and a half inches from wingtip to wingtip and, thanks to its helicopter-like hovering ability, can sneak up on spiders at the very heart of their webs and devour the succulent abdomens while the unfortunate victims are still alive. These are the closest living relatives of the science-fictional *Megasecoptera* that, judging from their fossilized remains, thrived in the swamps and forests of the Carboniferous era, more than three hundred million years ago. Gigantic creatures, with a wingspan of more than three feet, these resembled vampires with their grotesque sucking mouthparts, and, like those of the present-day dragonflies, their pond-dwelling larvae were ferocious predators; they grew up to twelve inches long.

Fossilized *Megasecoptera* had been known since 1885, and their remarkable forms would have been familiar to Wigglesworth. Equally familiar to him was the fact that dragonflies have been remarkably resistant to change during their long evolutionary existence, many still holding their wings outstretched at rest, though the more graceful damselflies have learned how to fold them together above their backs. The larvae of most dragonflies, though mere dwarfs in comparison to their ancestors, still grow to about two inches long. They are aquatic and come equipped with gills to breathe underwater, big eyes to help them spot their prey, such as insects and tadpoles, and an extensible lower jaw that shoots out and hooks their victims. And here, in the development of this ancient flying order we find another elementary clue to the evolution of metamorphosis. The dragonflies develop their wings on the outer surface of their bodies, from buds that are present during the larval stages and that grow at intervals throughout a series of molts.

Fossil nymphs of these ancient insects show scaled-down "winglets" that appear to have been articulated but held out stiffly to the side. These subsequently evolved to immovable pads folded over the back, protecting the developing wings from harm. This in turn ushered in the first "incomplete" form of metamorphosis, with the external pads we still see today in the larvae of certain insects. Meanwhile complete metamorphosis, such as we see today in butterflies and bees, evolved later, yet still more than three hundred million years ago. Some entomologists believe that there was a big expansion in the more colorful examples of complete metamorphosis, including moths and butterflies, at roughly the same time that the flowering plants were coming into the record, suggesting the possibility of an interlinked evolution, known as coevolution, in which we discover some of the loveliest jewels in the worlds of plants and insects.

What, Wigglesworth now asked himself, was the real difference between incomplete metamorphosis, such as we find in the dragonflies, and true or complete metamorphosis, such as we find in butterflies?

Of the twenty-eight orders that make up the class of the insects, the four most primitive orders molt but they do not metamorphose. Of the remaining twenty-four orders, fifteen undergo incomplete metamorphosis, the wings developing from buds on the exterior of the cuticle, with the buds gradually increasing in size with successive molts to complete their development with the final growth spurt. Besides the dragonflies, these external-bud orders include many familiar groups, such as mayflies, cockroaches, mantids, termites, earwigs, stick insects, and the grasshoppers, locusts, and crickets. In all of these we see a process of gradual growth, through molts, during development from larva to adult—albeit the greatest change is seen in the final molt. In these incompletely metamorphosing orders, there is no separate larval phase, such as the caterpillars or grubs of the fully metamorphosing orders, and no pupal stage before the final metamorphosis.

This contrasts with what is seen in the remaining nine orders, where the wings develop internally. The more specialized of these orders include the translucent-winged wasps, bees, and parasitic ichneumon flies, the double-winged true flies, the hundreds of thousands of species of beetles, and the quarter of a million species of butterflies and moths. In these last, the most varied and colorful of the insects, there is always a larval stage, often accompanied by voracious feeding and growth, and most telling of all, the transformation into the adult involves the bizarre process familiar from those earliest lessons at school, the same mystery that had beguiled naturalists since Aristotle, in which the caterpillar stops feeding and becomes cocooned from the world in a silky or leathery case, variously known as the pupa or chrysalis, within which some extraordinary changes in form and development take place, to result in the emergence of the fully formed adult insect.

11

The Phoenix in Its Crucible

Goliath beetle

When I visited Jonathan Wigglesworth, Vincent's son, he showed me a surviving display case from his father's personal collection of insects. I found myself gazing at an array of insects laid out like jewels, glittering and sheening, the majority of which I could never have put names to, yet all perfect in their sculptured forms. Prominent among them was a goliath beetle he might have brought home from his African safari. I couldn't help but wonder if, in that fabulous but slightly monstrous creature, he had considered the mystery of its birth through that striking transformation that had entranced him as a boy.

The African goliath beetle is the heaviest flying insect in the world. It belongs to the same family as the scarab, which was holy to the priests of ancient Egypt, and it can grow to 5 inches long and weigh up to 3.5 ounces. The female beetle lays her eggs, which are no more than a few millimeters in diameter, on the rich, moist, rotting detritus of the tropical forest floor. When the creamy caterpillar-like larvae hatch, they feed on dead wood, but as they grow bigger and stronger they become predators of other insects. This carnivorous diet is very nutritious, enabling them to grow so rapidly that within a few months they reach 6 inches in length. At this stage, they burrow into the forest floor and construct a protective cocoon. Here they pupate, undergoing a complete metamorphosis, before the gigantic adult beetle emerges fully grown, with its black-and-white striped thorax and its first pair of wings, evolved to hardened shields known as "elytra," which protect the flight wings—an apparatus so flexible and delicate, and yet so powerful, that when the Goliath takes to the wing it sounds like a miniature helicopter.

This extraordinary transformation within the pupa is one the goliath shares with the honeybee, the monarch butterfly, and Fabre's giant peacock—as well as with the entire order of beetles, or Coleoptera, with its 275,000 members. Take the larva of the goliath beetle—this is obviously a very different animal from the giant adult, much as the caterpillar is to the moth or butterfly, or the larval grub to the honeybee, wasp, or bluebottle fly. For the larva to metamorphose to the giant helicopter-buzzing adult, massive structural change must take place throughout the length, breadth, and internal structures of the larval body.

Rather than a den of repose, we see now that the enclosed chamber of the goliath's pupa really is a crucible tantamount to the mythic pyre of the phoenix, where the organic being is broken down into its primordial elements before being created anew. The immolation is not through flame but a voracious chemical digestion, yet the end result is much the same, with the emergence of the new being, equipped with complex wings, multifaceted compound eyes, and the many other changes necessary for its very different lifestyle and purpose.

The emerging adult needs an elaborate musculature to drive the wings. These muscles must be created anew since they are unlike any seen in the larva, and they demand a new respiratory system—in effect new lungs—to oxygenate them, with new breathing tubes, or tracheae, to feed their massive oxygen needs. The same high energy needs are supplied by changes in the structure of the heart, with a new nervous supply to drive the adult circulation, and a new blood to make that circulation work. We only have to consider the dramatic difference between a feeding grub or caterpillar and a flying butterfly or a beetle to grasp that the old mouth is rendered useless and must be replaced with new mouthparts, new salivary glands, new gut, new rectum. New legs must replace the creepy-crawly locomotion of the grub or caterpillar, and all must be clothed in a complex new skin, which in turn will manufacture the tough new external skeleton of the adult. Nowhere is the challenge of the new more demanding than in the nervous system—where a new brain is born. And no change is more practical to the new life-form than the newly constructed genitals essential for the most important new role of the adult form—the sexual reproduction of a new generation. The overwhelming destruction and reconstruction extends to the very cells that make up the individual tissues, where the larval tissues and organs are broken up and dissolved into an autodigested mush in which individual larval cells follow new developmental instructions, where, to all intents and purposes, life has returned to the embryonic state with the constituent cells in an undifferentiated form.

For us today, as for Vincent Wigglesworth three quarters of a century ago, this process is startling in its evolutionary implications. In Wigglesworth's own words, "These remarkable transformations raise questions which have exercised the minds of naturalists from the earliest times." And now he asked the pertinent questions: "How are the changes in form regulated? How did they arise in the course of evolution? What is the significance of the pupal stage?"[1]

In the 1930s, when Wigglesworth first began to explore this mystery, many of his colleagues believed that the answer to all three questions lay in structures known as "imaginal disks." The term *imaginal* has nothing to do with the imagination, but with the root word from which *imagination* itself springs: the Latin *imago*, which means "image" or "likeness." The imaginal disks are curious thickenings of the inner surface of the larval skin, or ectoderm. First described in 1762 by the Dutch naturalist and engraver Pieter Lyonet, whose monograph on the anatomy of the goat moth is regarded as one of the most beautifully illustrated works on anatomy ever published, imaginal disks were later found to occur only in insects that undergo full metamorphosis. Through careful study, biologists discovered that, during the pupal stage, the imaginal disks grow and transform into the complex structures of the adult, including the legs, the wings, the genital apparatus of male and female—and even the striking multifaceted compound eyes. In 1864, the German naturalist August Weismann proposed that the imaginal disks were actually nests of embryonic cells—what we today would call stem cells—containing the genetic blueprint for the future image of the adult insect. These lie dormant through the life of the larva only to be activated during the furious demolition and reconstruction within the pupa, when, on some key inner signal, they grow and develop into the specialized organs and tissues of the adult insect.

Wigglesworth had reservations about the purportedly unique role of the imaginal disks in metamorphosis. Interpretations of metamorphosis at the time centered around the more extreme examples, but in his words, "many parts of the adult body, even in such insects as Lepidoptera [moths and butterflies] or Hymenoptera [wasps, bees, and ants], which suffer a spectacular metamorphosis, are not formed from imaginal disks but are laid down by the same cells as have formed the larva." Moreover, even in the completely metamorphosing insects, there were exceptions to the wholesale immolation in the furnace of the pupa. For example, in some insects, such as the alder flies and scorpion flies, and even some of the less developed moths and butterflies, some larval muscles were retained to help create the abdomen of the adult. Even the superficial inertia and rigidity of the pupa was seen to vary. And in the pupa, the imaginal disks were not the sole source of adult tissues.

Such observations suggested that developmental mechanisms other than the imaginal disks were at work in the pupa—those same, or very similar, mechanisms that were at work in the less dramatic changes seen in the orders that underwent an incomplete pattern of metamorphosis.[2] All of this suggested to Wigglesworth that, perhaps, the dramatic changes seen in the pupa of butterflies and moths might not be as representative as the proponents of imaginal disks assumed. Taking this observation a stage further, he was encouraged to look for commonalities across all of metamorphosis, whether complete or incomplete.

Wigglesworth accepted, with reservation, the Darwinian viewpoint that the larva is subject to the influences of its environment. Through "natural selection (or whatever other agencies control organic change) it may undergo an evolution of its characters that leave the adult insect untouched."[3] It made perfect sense to Wigglesworth that natural selection was operating independently on the larval and adult stages, living as they did in their very different life cycles. It seemed a predictable sequel to this that as the differences between larvae and adults became more extreme, for example between the caterpillar and the butterfly or the maggot and the fly, that evolution would need to find some adaptive solution to the increasingly wide gap in body forms and physiology between the larval and adult insect. Here, to his mind, lay an important clue to the mystery of the pupa.

What if, putting the curious mystery of the pupa to one side, the key to metamorphosis was not confined to butterflies and moths, but extended to all orders of insects other than the primitive nonflying ones, whether their metamorphosis was complete or incomplete? To put it bluntly, what if there was no significant difference in the mechanisms of partial and complete metamorphosis?

If he was right on this crucial assumption, he could direct his researches at a single explanation that would satisfy all.

Aristotle taught that the embryonic life of insects merely continued until the formation of the perfect adult. "The larva while it is yet in growth," he wrote, "is a soft egg." But Wigglesworth disagreed. He took the view that the larva, pupa, and adult were a series of alternative forms a complex organism assumes at different periods of its development. The critical factor, for Wigglesworth, was what determined which form it assumed at any one stage.

He began to devote more and more of his time to his study of the great mystery. His day-to-day experimental methods are not revealed in his scientific publications, but it is possible to extrapolate somewhat from the recollections of those who worked with him.

Henry Bennet-Clark, today an emeritus reader in zoology at the University of Oxford, was a graduate student working under Wigglesworth's supervision at Cambridge from 1957 to 1960, where he studied molting in ticks and spiders before moving on to the mechanical properties of the expansile skin or "cuticle," an interest that would lead him to become a world expert in modeling insect movement.[4] In his years at Cambridge, he had plenty of opportunity to observe how Wigglesworth thought and worked. And his first, and most telling, observation was that Wigglesworth almost always worked alone. "This, I think, is reflected in his publications, which were something in the region of 250, of which I think only 17 had more than one author." In this he closely resembled another medical graduate and famous scientist, Jorgen Lehmann, who discovered the antituberculosis drug, PAS.[5]

"Why do you think he was so solitary in this?"

Suddenly our interview sprang alive. "Wigglesworth once said to me, 'Okay—we'd better arrange for you to get hold of some ticks.' And duly, some ticks appeared. I, now, thought—'Fine!' I was happy with this. Then he said, 'You'd better have a rabbit to feed them on.' I duly got a rabbit in the animal house. I bred up my ticks and started doing my experiments. After about six weeks, he said, 'Well—how's it going?' By then I'd learnt how to grow ticks. I'd started finding out some odd things about their biology, which put me onto the line of research I subsequently followed. I suppose he knew that I was there. He would come in, say, at quarter to nine every morning, and go out at quarter to one. Lunch at Caius. Come back at 2:00 p.m. Go home again at about five. He had his bicycle outside the lab somewhere. [Wigglesworth never learned to drive]. He cycled off to the northern part of Cambridge, where he lived. His daily routine was like that. I have no evidence whatsoever of what he did before a quarter to nine, or what he did after 5:00 p.m.

I suspect he, as it were, wrote *The Principles*, and so on. He was not what might be called an outstandingly clubbable man."

"But you must have talked to him about your work?"

"Every so often, when I thought I had got hold of something different—or any of us grad students thought we had something different—we'd go and see him and he'd sort of grunt."

"Was this in his office?"

"Either in his office or in his lab, where he would be bent over whatever he was doing to *Rhodnius*. And he would grunt. Every so often, when you'd actually done something good—he'd say, 'That's quite interesting.' He would sometimes say, 'Yes. I think that might be worth pursuing.'"

Other characteristics even more revealing than the air of preoccupied scientist ran through all of Wigglesworth's experiments, and Bennet-Clark captures this succinctly:

"First of all [his work], and particularly so his early papers, were immaculately illustrated. I think he must almost have thought as he was drawing: 'If I do this properly, it's going to be in books.' And they jolly well were, of course."

Another important characteristic of Wigglesworth's work was the small number of experimental subjects. As Bennet-Clark explained: "If you construct an experiment to test a prediction, and that prediction is borne out by results on five separate occasions, you can say, 'Whoopee! That is a one in thirty chance. That's a 3 percent probability. And I don't need to do it again.' If you read some of his papers you will actually find that it is quite obvious that the sample size was three. Wigglesworth, on several occasions, only did something once."[6]

At the same time Henry Bennet-Clark was working on the biomechanics of the insect cuticle, Simon Maddrell was working on urine production. Maddrell, who subsequently discovered the insect diuretic hormone, is now a fellow of the Royal Society, and holds the same professorship once occupied by Wigglesworth at Cambridge. He also recognized the unusual clarity of Wigglesworth's experimental thinking.[7] Such small sample sizes were not the mark of carelessness or laziness, as he explained. On the contrary, they implied deep thought and very careful planning of experiments in advance. In essence Wigglesworth formulated an experimental approach in which he required a specific answer to a question, yes or no. The experiment was constructed with such elegant simplicity and lucidity that very few test subjects were necessary.

We observe the same clarity of approach when Wigglesworth now perceived that the physiological factors controlling metamorphosis could be encapsulated in two simple questions. What factors inhibit the development of the imaginal disks in the young stages? And what factors cause the larval tissues to disintegrate and activate the imaginal buds at metamorphosis?

If he was right in ignoring the widely held views of critical differences between complete and incomplete metamorphosis, then the answers to these two questions would prove to be the same in all examples of metamorphosis. He could assume that the latency and subsequent realization of the adult insect can be learned just as well from incompletely metamorphosing insects, in which imaginal disks do not occur. Conveniently, he was already familiar with an insect that fitted the bill, an experimental subject that was tough and hardy, easy to keep alive in the underground insectaries of the London School—an insect he had already found amenable to clear and simple experiment.

In his words: "It was on these grounds that *Rhodnius prolixus* was chosen for the study of metamorphosis."

— 12 —

Two Souls in One Body

Kissing bug

Rhodnius prolixus, otherwise known as the kissing bug, is a parasitic insect of the drier savannah areas, extending from southern Mexico to northern South America. It is a member of the order Hemiptera, commonly known as "bugs," many of which, as Wigglesworth himself expressed it, "interfere seriously in the affairs of man." The Hemiptera include more than fifty-five thousand species, most of which, such as cicadas, aphids, and greenflies, have penetrating mouthparts that enable them to feed on the sap of plants. Others use those same syringelike mouths to suck the body fluids of other insects or, like *Rhodnius,* to suck the blood of mammals and birds.

During its life cycle the wings develop outside the body, progressively enlarging during successive molts from externally placed "lobes" or "pads." There is no dramatically different larval phase or life cycle, such as we see in the caterpillars of butterflies and moths, and no pupal stage. Instead the insect grows and develops through a series of molts, completing growth and development during the final molt. In this way *Rhodnius* is typical of the insects that undergo incomplete metamorphosis. Sap- and blood-sucking insects are notorious for transmitting disease-causing microbes, particularly viruses, from plant to plant or from animal to animal, and *Rhodnius prolixus* is no exception. Fairly big for a disease-transmitting insect, the fully grown *Rhodnius* is about an inch long, a fact I was privileged to witness when a medical entomologist, Professor Chris Curtis, let me handle living specimens in the underground insectaries at the London School of Hygiene and Tropical Medicine. Its body is shaped like an extended prism, with a long narrow cylindrical head and an abdomen capable of enormous syringelike expansion, an adaptation designed to accommodate its predatory lifestyle. The larval stages of *Rhodnius* ingest from six to twelve times their own weight of blood at a single meal in preparation for molting. In explaining why he chose this insect for his study of metamorphosis, and many subsequent investigations, Wigglesworth freely declared: "It so happens that in *Rhodnius*, this single meal of blood suffices for each [larval stage], and that molting occurs at a definite interval after this meal, the whole process following a more or less constant timetable. This circumstance makes *Rhodnius* an admirable subject for such an investigation."[1]

In its natural habitat, *Rhodnius* inhabits a number of ecologies, including the crowns of palm trees, from where it swoops to feed on the blood of marsupials, rats, reptiles, carrion birds—and humans. It is the insect vector of the parasite *Trypanosoma cruzi*, which causes an extremely unpleasant illness known as Chagas disease, a condition that is largely incurable, afflicting the intestines, heart, and nervous system. Some authorities believe that this was the real nature of the illness that plagued Charles Darwin for decades of his life and that eventually caused his death.[2]

This insect now took central stage in Wigglesworth's intensive exploration of metamorphosis. And like the great naturalists before him, he began by making full use of intelligent observation.

The larval stages of an incompletely metamorphosing insect are known as nymphs, and the molting from one larval stage to another is known as "ecdysis." *Rhodnius* molts through five nymphal stages, during which the structures and pigment patterns of the surface cuticle show little change. There is progressive increase in size and development of the wing lobes and, in the later stages, the genitalia begin to appear. But these changes are slight compared to the major development that occurs in the final molt, when the fifth nymph changes into the adult. Elaborate genitals become fully developed, the thorax, or chest, adopts its final size and shape, strengthened to accommodate the flight muscles, and the wing lobes are massively transformed into the large and powerful wings. Compound multi-faceted eyes now appear for the first time. Meanwhile, the structure and coloring pattern of the cuticle is dramatically changed.

For Wigglesworth, who was convinced that incomplete and complete metamorphosis could be considered as essentially the same mystery, "It is therefore convenient . . . to refer to this final stage as 'metamorphosis.'"[3] Now, examining the changes that took place during metamorphosis, it was clear that these changes embraced the entire body of the insect, including skin, eyes, muscles, sexual organs, and wings. For such universal changes to happen abruptly, during the final molt of an insect such as *Rhodnius*, there had to be some form of a message, or signal, capable of being transmitted to every tissue and organ throughout the insect body. Very few systems were capable of bodywide transmission in this way. Wigglesworth already knew, from the work of other biologists, that the wing muscles, in their powerful development during the final molt, required an intact central nervous system. And the nervous system was capable of transmitting a message or signal bodywide. Now he asked himself the obvious question: was it possible that the nervous system held the key to metamorphosis?

The European gypsy moth, *Lymantria dispar*, is a member of the tussock family and a considerable pest in forestry. The adult female is pretty, if decidedly hairy, and its creamy wings, dotted with dark spots, extend to about two inches in span. It is too large and heavy to fly; instead it hauls itself up onto the trunks of trees and shrubs, where it exudes a pheromone to attract the smaller and browner male. Once mated, it lays a single egg mass, after which the female dies, allowing up to a thousand caterpillars to emerge at the end of April or in early May. These are both ravenous and omnivorous in their tastes. Such is their hunger, they will defoliate more than 450 different species of plants, but oaks and alders are their preferred broad-leaved diet and Douglas Fir and Western Hemlock their needle-leaved alternatives. When the infestation is extensive, the result, in arboreal terms, is an epidemic. In 1923, Stefan Kopeč, professor of biology at Warsaw University, chose the gypsy moth as the subject of a series of experiments aimed at examining the role of the nervous system in metamorphosis.

In a key experiment, he removed the nerve centers known as the thoracic ganglia from pupae and waited to see how this would affect their subsequent metamorphosis. The wing muscles failed to develop. But it was not critical, as Kopeč went further to show that growth of the body form of the adult, which is derived from the epidermis, was otherwise normal.[4] Another scientist showed that if you remove the corresponding nerve ganglia in the praying mantis, *Sphodromantis*, the antennae and legs still grow to normal size and shape—in spite of the fact they contain no nerves or muscles. Wigglesworth got the message—the nervous system was not the ultimate key to metamorphosis—and he switched his focus to the skin.

The insect epidermis fulfils many different roles, and different areas of skin in the same adult insect will have different appearances and properties. When he excised a small area of the skin of the fourth-stage nymph of *Rhodnius* and implanted it into a different area of the body surface of another fourth-stage nymph, the transplanted skin healed into its new site and appeared little different from the rest of the skin. But when the recipient nymph metamorphosed to an adult, the implanted skin developed the adult pattern and structure of the part of the body he had taken it from in the first place. Through these and other examples, Wigglesworth proved that even in insects that underwent incomplete metamorphosis there were major changes in epidermal structure as part of the development from the nymph to the adult.

He sat back and thought about it all again. It was clear that the fertilized egg contained the hereditary programming for two very different and specialized patterns of body development, one larval and one adult. But the expression of the adult program was intrinsically different in the two major forms of metamorphosis. In complete metamorphosis, the adult features arose from stem cells that had been set aside for the pupal stage in the imaginal disks; in incomplete metamorphosis, such as we see in *Rhodnius*, the adult body form was present, as a genetic or developmental blueprint, yet latent, throughout the larval cells at the time when they are expressing the nymphal pattern of development.

This led him to pose two new questions. Some mysterious organ in the insect body must control this curious pattern of duplex development. What organ was it? And how did it control development?

Wigglesworth reminded himself of the fact that metamorphosis from the insect larva to the adult, whether from nymph direct to adult, or from caterpillar, through the intermediate stage of pupa, to adult, affected every cell in the insect's body at one and the same time. Given that the nervous system had been excluded as the controlling influence, the only other system capable of broadcasting a signal in such a universal manner was the insect bloodstream.

In 1922, Kopeč appeared to have confirmed this in the development of the gypsy moth.[5] For Wigglesworth, the most likely explanation was that the signal for metamorphosis was a blood-borne chemical signal—what today we would call a "hormone."

— 13 —

Bizarre Extrapolations

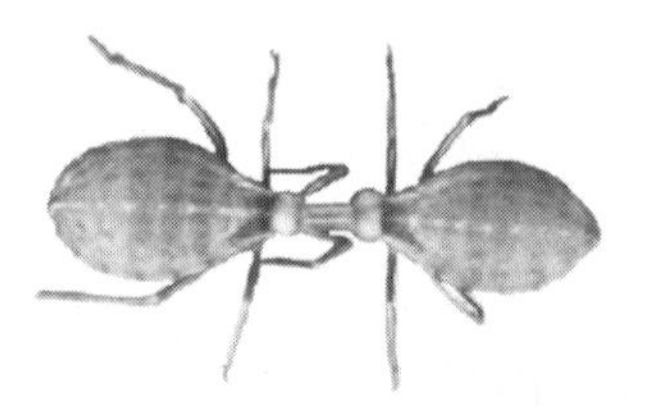

Parabiotically conjoined insects

Goliath, in the Book of David, stood six cubits and a span—in modern terms 9 feet 9 inches tall—while Tom Thumb, in traditional English folklore, was said to be small enough to hitch a ride on a butterfly. In actual recorded history, the tallest man was Robert Pershing Wadlow, a native of Alton, Illinois, who was 8 feet 11.1 inches tall, while the tallest woman was the Chinese Zen Jinlian, who grew to be 8 feet 1.75 inches. Up to the end of the nineteenth century, we had no means of knowing how such extremes of stature came to be. Yet in the lives of ordinary men and women sudden growth, and wholesale changes in body size and shape, and changes in internal development, are our common experience as part of the strange, often disquieting, events of puberty.

Before puberty, boys and girls have the same proportion of muscle and skeletal mass and body fat. However, during puberty, boys and girls undergo dramatic physical development, including rapid body growth, and major changes in muscle and fat distribution that differ between the sexes. So profound are these changes that, by the end of puberty, healthy males have 1.5 times the proportional skeletal and muscle mass of females, whereas the healthy females now possess twice as much proportional body fat as males. These visible changes are accompanied by major cellular and tissue changes involving sexual and related organs, such as the breasts and genital organs in females and the genital organs in males. Today we know that this involves a second phase of activation of developmental genetic pathways—the same sort of genetic pathways that are normally involved in embryonic development in the womb, so that, in the scientific jargon, human puberty is a phase of post-embryonic development. Perhaps more interesting still, recent research suggests that the processes of metamorphosis and puberty have so many commonalities that some developmental biologists suggest that "puberty may be considered a variation of the metamorphic theme."[1]

With the discovery of the human gonad-stimulating hormone and other human pituitary hormones in the 1920s and 1930s, hormones became a buzzword in medical and mammalian animal research.[2] Tiny secretions that circulated internally from specialist "endocrine" glands, these played a vital role in normal body development and health. For example, the thyroid hormone, secreted by the thyroid gland in mammals, was found to control the maturation and speed of the body's physiology and chemistry. Puberty was brought about in humans by pulses of gonadotropin-releasing hormone, which was released by the hypothalamic portion of the brain. This stimulated the master gland, known as the pituitary, to increase the secretion and release of the sex-gland-stimulating hormones, or gonadotropins, which traveled through the bloodstream to the ovaries and testes, where they provoked increasing blood levels of estrogens and androgens. These hormones determined whether the body developed as a male or female. All of this would have been familiar to the medically qualified Wigglesworth at the time he was conducting the metamorphosis research. But there was a major problem in extrapolating the vertebrate hormone discoveries to insect metamorphosis: earlier experiments appeared to have proved that insects did not possess hormones. In Poland, Kopeč had gone so far as to propose that his results might indeed indicate the presence of a hormonal signal, but general opinion had accepted that insects had no hormones and so Kopeč's proposal had been largely ignored.[3]

Wigglesworth was aware that Kopeč had shown that the insect brain appeared to be the source of the initial stimulus for molting. When, in what must appear a bizarre experiment, Kopeč beheaded the gypsy moth caterpillar, the brainless larva didn't necessarily die, but the subsequent molt to the pupa failed. Another colleague had demonstrated that if a severed brain was reimplanted into the headless caterpillar of the elephant hawk moth, *Deilephila*, it restored the ability to pupate. Much the same had been shown in the caterpillar of the agricultural pest moth, *Ephestia*, and the silkworm moth, *Bombyx mori*. What then was the real nature of the metamorphosis control factor? Wigglesworth was increasingly convinced that Kopeč had been right all along in suggesting it was a hormone. And, if it truly was a hormone, the key question was, where was it being manufactured and released?

Wigglesworth explored this further in some bizarre experiments that are now considered classics in insect physiology.

In his preliminary paper, in 1933, he had described in detail the wide array of physiological changes of metamorphosis in *Rhodnius prolixus*. In 1934, in a second lengthy paper, he searched for the site of production of the hormone that he presumed was responsible for such large-scale changes.[4] He began by reminding his readers that *Rhodnius* invariably goes through five nymphal stages. In each of the first four molts it maintains the same body structures. Only during the fifth molt, when the insect becomes adult, does it go through the striking changes of its metamorphosis. As Kopeč had already demonstrated, beheading an insect larva did not necessarily kill it, but this drastic action had had a profound effect on future development. To check this out for himself, Wigglesworth now beheaded examples of all five stages of *Rhodnius* larvae, using a ligature of thread in the larger forms and scissors in the smaller ones. The wounds caused by the beheading were sealed with molten paraffin wax.

In each stage, he confirmed Kopeč's 1922 observation that there was a critical time period in relation to the blood meal that determined the developmental effects of beheading. If he beheaded the nymph before this critical time, the molt did not take place, and if he beheaded it after this time it did. In the fifth *Rhodnius* molt, which normally took place about twenty-eight days after a blood meal—it seems likely that at this stage in his experimentation Wigglesworth himself was the provider of the blood—he found that he could pin down this critical time to between the sixth and eighth day. When he beheaded the insect larvae after the eighth day, the headless insects continued to digest their blood meals, excreted their waste and went on to molt, the process following the normal timing and pattern. The insects beheaded before these dates, though they survived for vast periods, even as long as a year, failed to molt.

For Wigglesworth, this confirmed Kopeč's 1922 observation. He was convinced that a hormone was involved and that it was released in or near to the head. The larvae decapitated early had lost the hormone but in those decapitated close to the onset of molting, the successful molt suggested that the hormone had already been secreted and carried via the blood to the remainder of the body. Wigglesworth was ready to design a new experiment that would further test this hypothesis. In his own words, "If this idea is correct, the blood from an insect that has just passed the critical period should induce molting in an insect decapitated before that period."

The experiments became ever more bizarre. He beheaded two groups of nymphs, one (the primary) shortly after the critical period had passed and the other (the recipient) just twenty-four hours after feeding, and thus not expected to molt on their own. Then he used paraffin wax to join the two insect bodies together, neck-to-neck, so that the blood could flow freely from one to the other—an experimental technique he called "parabiosis." Wigglesworth found that a number of the primary group molted and as they did so they induced molting in the recipients. He went on to repeat the experiment many times with nymphs at different stages. In every case where the primary insect was decapitated soon after the critical watershed period, its blood induced molting in a recipient decapitated twenty-four hours after feeding.

The conclusion was clear: molting was indeed initiated by a blood-borne factor—almost certainly a hormone.

But he also spotted something else, an observation that now seems precocious given the relative ignorance of developmental biology at that time. When he monitored the exact timing of metamorphosis in the beheaded and surgically conjoined insects, the onset of molting was accelerated in the recipient insect—it would take six days, for example, to reach a stage normally reached in nine or ten days—while molting in the primary insect appeared to be delayed. These curious accelerations and delays lasted until the recipient insect had caught up with the development of the primary. It was as if the two insects were being integrated into one body, with different parts of the conjoined body being brought into developmental alignment. This led Wigglesworth to a brilliant deduction:

"It is inconceivable that a single external factor [i.e., hormone] should ensure this exact coordination. These observations can only mean that the growing tissues themselves are communicating with and controlling one another by chemical means." As if grasping the future importance of this observation, he added, "This is a fundamental idea but it is beyond the scope of the present work."[5] He had anticipated the discovery of signaling from cell to cell and from tissue to tissue, which would only come to be fully appreciated some half a century later by the new science of evolutionary development, or "evo-devo."

In fact, what he was now observing might well have homologies with the groundbreaking discoveries being made in human hormonal research. Wigglesworth went on to suggest that the "molting hormone" could only be secreted "in the head or, more likely, by some organ innervated by nerves coming from the head." But what possible organ could be the source? He focused on an organ known as the corpus allatum—or more correctly the corpora allata, since these existed as paired structures on either side of the insect esophagus, within the head. Named by the French biologist

C. H. Janet in 1899, the function of the corpus allatum was entirely unknown. Like the endocrine glands of vertebrates, it budded off from the epidermis at an early stage of development, and it was well supplied with nerves. Indeed there were many pointers to this being the insect equivalent of an endocrine gland, which appeared to have close connections with a major nerve ganglion and through it with the brain. Wigglesworth began a series of careful dissection and staining studies, confirming that the corpus allatum was the only organ in the head that showed a definite cycle of changes in relation to the critical period after larval feeding. In doing so he also made an observation that seemed both intriguing and baffling. During the fourth nymphal molt, he observed a major increase in the number of the cells in the corpus allatum, with cell divisions, known as mitoses, being very numerous during the third, fourth, and fifth days after feeding. But during the final molt of the fifth nymph, the molt that gave rise to *Rhodnius*'s metamorphosis, he saw no such increase in cells. On the contrary, the gland showed no increase in size and no cellular mitoses were seen. To Wigglesworth, this suggested that the corpus allatum was active in normal molts and was then switched off, for some reason, during the final molt to the adult. The corpus allatum had to be the source of the molting hormone.

Wigglesworth had also observed that molting always followed a blood meal, yet did not occur if the nymph was supplied with a steady trickle of blood instead of a single feast. This suggested that it wasn't the state of nutrition of the nymph but the massive abdominal distension of the blood feast that triggered the secretion of the molting hormone. Further experiment now confirmed that this was the case. From this he deduced that the stretching of the abdomen sent a signal to the brain, which in turn signaled the corpus allatum to get to work. This too he tested by cutting the nerve cord immediately behind the neck twenty-four hours after a full meal. Of thirty-two nymphs treated in this way, only six molted.

Wigglesworth sat back and considered his options. What then was the mechanism by which the fifth and final molt led not to a larger nymph but to the markedly different adult form?

There were two possible explanations. One possibility was that the hormonal control in the fifth molt was identical to that in all the other molts but the body cells at this stage had developed some inherent difference, so they responded differently to the same hormonal stimulus. The other possibility was a change in the actual hormones themselves. He tested these alternatives with a new experiment, joining a fifth-stage nymph that had been decapitated after the critical period to a fourth-stage nymph decapitated twenty-four hours after feeding. Both insects molted simultaneously. The smaller, fourth-stage nymph suffered a premature metamorphosis into a smaller but otherwise normal adult. He repeated the procedure with first-stage nymphs, which would normally molt to a larger, second-stage nymph. Instead they metamorphosed to diminutive adults. He could draw two possible conclusions. Either the molting hormone of the final nymph was different from that released in the earlier molts; or the hormone was always the same but the earlier nymphs also produced an inhibitory hormone which was absent in the final molt to the adult form.

He moved on to another experiment, joining fourth-stage nymphs reared to beyond the watershed for hormone production to fifth-stage nymphs decapitated twenty-four hours after feeding. The fourth-stage nymphs, when they molted, induced molting in conjoined fifth-stage nymphs. But the fifth-stage nymphs did not molt to adults. They retained the larval characteristics. Wigglesworth had his answer: "The molting hormone is the same at metamorphosis as at earlier molts. Therefore the absence of metamorphosis in the younger insects must be due to an inhibitory factor."[6] This was an important breakthrough.

Wigglesworth had no idea where this inhibitory factor was produced, but one possibility was that it was secreted in the head. He even surmised that the corpus allatum might be the source of both the molting and inhibitory hormones, through different types of cells stimulated as a result of different programs controlled by the brain.[7] Time would prove that, though he was spectacularly perceptive in his many experimental observations, in this general conclusion of two hormones being secreted by the corpus allatum he was mistaken, at least in its universal application.

A decade later, in his book *The Physiology of Insect Metamorphosis*, he would chide himself for his error, which had been made "without good experimental evidence." His conclusion was disproved by studies conducted in moths by French and German colleagues, J-J. Bounhiol and E. Plagge, respectively. These experiments proved beyond doubt that the corpus allatum was the source of some important secretion—but it was not the source of the molting hormone.

Wigglesworth extended his experiments to include a wide diversity of insects, including dragonflies, mayflies, locusts, and grasshoppers in addition to butterflies, moths, and, of course, the blood-sucking *Rhodnius*. Heads were joined together, resulting in double-bodied and double-headed parabiotic conglomerates. Headless bodies were joined to intact heads of other bodies, resulting in conglomerates that had two bodies and one head. And, finally, this frenzy of arcane experimentation gave him his most important clues to answering the mystery.

The first clue came when he noticed that if larvae of *Rhodnius* in any nymphal stage, from first to fourth, were decapitated soon *after* the critical period, they tended to undergo a precocious metamorphosis that led to adult characters in varying degree. The presence of an intact head was important. "There is clearly some influence exerted by the head that is preventing the realization of imaginal [adult] characters latent within the cells of the larva."

Wigglesworth demonstrated that if he joined the headless body of a first-stage *Rhodnius* to the head of a fifth-stage larva in the process of molting, the tiny first-stage metamorphosed to a precocious adult. A hormone was clearly involved, but it couldn't simply be the molting hormone—otherwise the first-stage larva would have molted not to an adult but to a second stage. The penny dropped when Wigglesworth realized that what he was witnessing was not a stimulation effect but one of inhibition. If the corpora allata were removed from young silkworm larvae, they proceeded to spin cocoons and to pupate into miniature adults. Much the same thing was shown with the wax moth, *Galleria*, in which removal of the corpora allata gives rise to tiny pupae and miniature adults. This did not happen if corpora allata from other larvae were transplanted into them. Conversely, if corpora allata were transplanted into full-grown larvae ready to metamorphose into the adult moths, metamorphosis was canceled and larval development continued until it eventually produced giant pupae and giant adults. On the other hand, the application of a ligature behind the head in the larvae of the honeybee, at the right time, caused them to molt directly to forms showing adult characters, including hair, skin, and the honey-gathering brush on the hind feet. The corpus allatum certainly was producing a hormone, but its action was not to induce metamorphosis but to prevent the final change to the adult form.

Now those observations of a virtual closing down of the corpus allatum in anticipation of the final molt made more sense. In choosing a fifth-stage larva that was already entering metamorphosis, he had selected it at a stage when the inhibiting hormone had been turned off to allow the final metamorphic change. The first-stage larva would normally have produced this inhibitory hormone for itself and thus prevented the metamorphosis to adult at this precocious stage. But during decapitation it too had lost its supply of the inhibiting hormone, so the same hormone crossing over into its circulation from the metamorphosing fifth nymph had triggered the precocious change.

In a further major paper published in January 1940, Wigglesworth stated with certainty that the inhibitory hormone "is secreted by the corpus allatum in the first four nymphal stages."[8] This inhibitory effect was shown to cross species, genera, and even families of insect. For example, the corpora allata of the tobacco moth, the domesticated silk moth, the mealworm beetle, and the stick insect were variously shown to inhibit larval to adult metamorphosis in the wax moth. In this same paper he decided he could no longer call his discovery the inhibitory hormone. "In previous papers the 'inhibitory hormone' was so called because in its presence the production of imaginal [adult] characters at molting is suppressed. But now, in view of its probable mode of action through the activation of the nymphal system at the expense of the [adult], it might be preferable to refer to this hormone as the 'nymphal' or 'juvenile' hormone."[9] The world of biology accepted his argument, and the hormone is now known as the juvenile hormone.

Wigglesworth could now itemize the series of events that gave rise to insect metamorphosis. Stretching of the abdomen of the nymph set up a nerve reflex that traveled to the brain, very likely to a region of the brain identified as the "pars intercerebralis," only recently discovered by Bounhiol and Plagge.[10] Where he had earlier concluded that the brain now sent a signal to the corpus allatum, which then secreted the molting hormone, he now admitted that he had been mistaken.[11] The molting hormone came from some other, as yet, unknown source, a mystery that had proved to be a considerable source of irritation to Wigglesworth, but which he was now obliged to leave to others to resolve.

In 1931, the German entomologist V. Hachlow had shown that pupal development in the red admiral and the black-veined white butterflies was initiated by some center located in the thorax.[12] In 1940, a Japanese biologist, S. Fukuda, identified the structure responsible: a diffuse organ made up of beadlike strings of cells wrapped around the trachea in the thoracic segment of caterpillars.[13] It was all the more ironic that this gland had been described as long ago as 1762 by the Dutch naturalist and engraver Pieter Lyonet. In fact a number of naturalists had identified the same structure in different insects, in particular two more Japanese colleagues, K. Toyama, who had rediscovered it in 1902, and O. Ke, who, in 1930, had actually named it the "prothoracic gland."[14] In a series of elegant studies in the silkworm moth, Fukuda now demonstrated that if the larva was ligated behind the prothoracic gland, the front end pupated normally and the hind part did not. But if he transplanted some prothoracic gland into the hind part after ligation, pupation was restored. At some critical period, the prothoracic gland released an "active principle" into the blood that controlled the molting of the larva and the development of the pupa.

Perhaps we should pause to acknowledge how amazing it is that, at the very heart of World War Two, scientists in Britain, France, Germany, and Japan continued to engage in the experimental exploration of the great mystery of metamorphosis in insects.

On the night of Saturday, May 10, 1941, at about 10:45 p.m., the London School of Hygiene and Tropical Medicine took a direct hit with a bomb. Nobody was killed, though there were some people working there at this late hour. A sixth of the building was destroyed, partly by explosion and partly by fire "as it was not till morning that a fire brigade could be spared to come." The basement of the Mallet wing was severely damaged, as was the furniture in several other departments. The devastation was so great its effects were still visible into the 1960s,[15] but there was no interruption to Wigglesworth's experiments. At the outbreak of war, anticipating such difficulties, he had taken the precaution of transferring much of his lab and equipment to his home in Beaconsfield, ten miles or so from the outer limits of London.[16] How difficult it was for the many other scientists working in war-torn countries, one can scarcely imagine. For Wigglesworth, dividing his work between the London School and home, the situation had become a good deal more complicated than he had originally envisaged, with three separate organs identified as playing some role in metamorphosis, the pars intercerebralis of the brain, the corpus allatum, and the prothoracic gland. It now seemed inescapable that metamorphosis was not the result of a single active principle but a succession of them. All that remained was to bring all three together into a cohesive overall explanation.

It was at this point that, in the words of Bennet-Clark, "Then there was a really awful blow to Wigglesworth's life. You've come across Carroll Williams?"

"Yes," I replied.

"Carroll Williams represented, if I may say so, a major blow to Wigglesworth because he solved the puzzle of the molting hormone."[17]

The search for answers had now extended to America.

— 14 —

Assembling the Jigsaw Puzzle

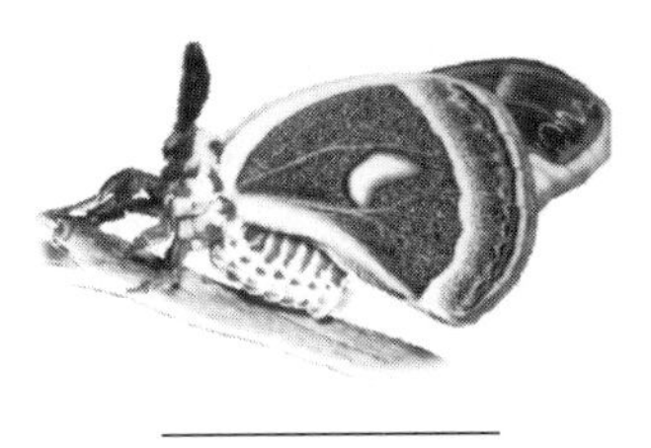

Giant silkworm moth

Carroll Milton Williams was born in Richmond, Virginia, on December 2, 1916. Like Wigglesworth he became entranced with insects in his youth, so much so that during his undergraduate years at the University of Richmond, he was appointed Curator in Insects to the university museum, and helped the state entomologist of Virginia to identify thousands of local specimens brought in by the public. One species of butterfly took his fancy. This was the magnificent Diana fritillary, which is somewhat rare in America, being found only in a narrow geographic range extending from West Virginia to Missouri. Unlike other fritillaries, its large black-and-blue females are utterly different in color from the smaller golden-orange and ultramarine males, so much so that only a skilled naturalist would even guess that they are related. This led to Williams's first professional publication, a description of *Argynnis diana* and other butterflies native to Virginia, in 1937, when he was twenty and just prior to graduation.[1]

That autumn he enrolled as a graduate student at Harvard, where, with the help of fellow graduate student Leigh Chadwick and advice from the ingenious Harold "Doc" Edgerton of MIT, he designed a stroboscopic apparatus that enabled him to measure the wing beat of the fruit fly, which registered at between 12,000 and 14,000 times a minute, a prodigious feat of muscular work ethic that could be sustained for as much as three hours or two million double beats. "Finally," in the words of A. M. Pappenheimer Jr., "he succeeded in demonstrating, where others had failed, the neuromuscular network in the thorax that controls the wing beat."[2] Williams completed what was described as a "remarkable and brilliant" analysis for his PhD, granted in 1941, on "A Morphological and Physiological Analysis of the Flight of Drosophila."[3] World War Two intervened. Having now joined the army, he was sent to Harvard Medical School "to tool up on tropical medicine" and went on to gain an MD summa cum laude. "Meanwhile," as he would later recount to a reporter from the *Boston Sunday Herald*, "I had begun to work on a part-time basis on the physiology of insect metamorphosis."[4] In 1946, with the end of World War Two, and recently appointed assistant professor of zoology at Harvard, he returned to his insect studies.

To contribute to the ongoing research into insect metamorphosis, Williams needed a larger test animal than the tiny fruit fly. He decided on the giant silkworm moth, *Hyalophora cecropia*. The moth, which is North America's largest, is strikingly beautiful, its body cloaked in cardinal red, topped off with an ermine-white collar and its head crowned with huge feathery antennae. The velvety wings are a mosaic of patterned reds, browns, grays, and old gold, with flamboyant crescents and two arresting eyespots, the wingspan extending in the largest specimens to something close to six inches. Like Fabre's great peacock, the adult moth does not feed. Its sole purpose is reproduction. The female lays her neat rows of eggs on the leaves of sandbar willows and other host plants during early summer, and these hatch as tiny caterpillars of a waxy green color. Feeding on trees and shrubs, such as the box elder and sugar maple—they will even devour the leaves of wild cherries, plums, and apples—the caterpillars must eat voraciously to grow to a prodigious four inches long and an inch wide.

At this stage they are ready to spin their silken cocoons, which are attached along their length to a twig, hidden away in some dark and protected area. In many species of butterfly or moth, pupation and the final metamorphosis to the adult begins soon after the caterpillar enters the cocoon. But, critically for Williams's research, the cecropia moth follows a different strategy. When the pupa is sealed within its cocoon, it immediately enters a period of quiescence, known as "diapause," a process akin to hibernation in vertebrate animals, that tides the insect over the following winter.

The change from activity to diapause happens abruptly, as if an internal switch has been thrown. At the same time all growth and differentiation comes to a standstill. But come the spring, when diapause ends, it's as if a second switch is thrown, and this ends the quiescence and kick-starts the furious activity of destruction and reorganization within the pupa that leads to the metamorphosis into the adult moth.

These dramatic changes, associated with diapause, fascinated the thirty-one-year-old biologist.

When Williams began his research, nobody knew the nature of the two mysterious switches, or what threw them first to quiescence, then to dramatic metamorphic change. "Whatever may be the inner mechanism for the induction and termination of diapause," he remarked, "it must have the capacity to turn morphogenesis off and on in a most striking way." This had so intrigued biologists that, in 1932, a French colleague named G. Cousin cited 347 papers on this subject in a single review of the literature.[5] Now Carroll M. Williams was determined to solve it.

He began by establishing that a spell of cold weather was critical for the metamorphosis of his cecropia pupae. In nature this was provided by winter. But if he manufactured an artificial winter, say by chilling the pupae at temperatures of 37°F to 41°F (3°C to 5°C) for a few weeks, adult development could be brought on by raising the temperature. Now that he knew that temperature increase threw the second switch, it afforded him a means of investigating the switch itself. He invented a way of anaesthetizing his insects with carbon dioxide, which was not only more humane but also permitted extensive and prolonged surgical manipulations without loss of blood or apparent damage to the pupae.[6] By 1946 Williams had performed 690 studies in 1,200 pupae, leading to the first of a series of papers in which he searched for the factors controlling diapause.[7] In this opening paper he laid bare his thinking: "If the termination of diapause is indeed accomplished by the action, within the previously chilled pupa, of a factor necessary for adult development, then it should be possible to demonstrate the organ in which this factor arises."[8]

When he removed the brain from a pupa, he found that this led to a complete failure of metamorphosis. The larva persisted in suspended animation until death, which could take up to two years. But if he took the brain of a chilled pupa and placed it in the abdomen of a headless resting pupa at 77°F (25°C), the isolated abdomen metamorphosed to the adult equivalent. It didn't seem to matter that the recipient was headless any more than it mattered if the abdomen belonged to a different species of insect, such as the American saturniid silk moth, *Antherea polyphemus*. In macabre fashion, the isolated abdomens would molt, attract a mate, and lay eggs. Non-chilled brains would not work. Thus he concluded "only one organ in the chilled pupa has the power to evoke development . . . this organ is the brain itself."[9]

The discovery that the insect brain controlled diapause fitted with Wigglesworth's findings of the importance of the brain during molting in *Rhodnius*. It also fitted with the environmental cue Wigglesworth had established: while the brain of *Rhodnius* responded to the signal of larval abdominal distension, the cecropia moth brain responded to an increase in ambient temperature. These results also fitted with the findings of other biologists who had shown that the brain was equally important during the pupation of completely metamorphosing butterflies and moths. Williams concluded his paper with the teaser: "What then is the nature of the factor arising in the brain that so spectacularly evokes the changes of metamorphosis in the chilled pupae?"

One possibility was that the implanted brain worked by some direct effect on the host tissues, converting them from dormancy to activity. But there was another, more intriguing possibility: the transplanted brain—perhaps the pars intercerebralis discovered by Bounhiol and Plagge—might act indirectly, by producing a hormone that stimulated a more peripheral organ in the recipient larva to produce the molting hormone. To evaluate this, Williams needed to chop up individual pupae into smaller fractions. But he knew from experience that division of the much larger silk moths by ligatures, such as pioneered by Kopeč, had not met with success. Even less appealing was the technique, pioneered in experiments on other insects, of slicing the unfortunate pupae into traumatized parts. Such brutal techniques were condemned to failure because chopping through the midgut filled the blood with lumps of tissue, and this killed the test animals through embolizing the heart with solid debris. He set about inventing a less brutal method, carried out under carbon dioxide anesthesia, in which he carefully dissected out and separated the lower six abdominal segments of brainless diapausing pupae from the front end of the animal, meanwhile keeping the midgut intact. The gut could be brought forward into the front end or dissected out and removed entirely. This led to a controlled severance of the front from the rear half of the pupae. He then sealed the cut ends with paraffin wax, in which he embedded a microscopic cover slip. The cover slip had a central aperture through which he could inject Ringer's solution, so as to maintain physiological homeostasis and keep out air. The hole, which was sealed with more paraffin wax, could be reopened at any future stage to allow him to operate through it: in the meanwhile the cover slip allowed him a window for inspection.

Despite his anesthesia and skillful dissection, two-thirds of the isolated pupal parts died within days. But the remainder survived for lengthy periods. Into each of the separated halves of the pupae he now implanted a chilled cecropia brain. The front ends metamorphosed normally to the corresponding front end of a lively moth. Meanwhile, the isolated abdomens remained undeveloped. He found that he could transplant as many as six chilled brains into a single abdomen without inducing a jot of change, even though these abdomens survived for as long as eight months. In his words, "This difference in response might be explained if the anterior fragment possessed a second developmental center that was lacking in the posterior fragment."[10]

More experiments followed. Isolated abdomens were grafted to brainless pupae, or to the front end of the severed bodies of brainless pupae. Graft and host joined up to grow together, but they failed to metamorphose. But if he introduced a chilled brain to either combination, the entire parabiotic construction metamorphosed to the corresponding parts of the adult. He moved on to grafting individual internal organs. "This search was vastly aided by the publication of Fukuda's paper in 1941," as Williams would subsequently acknowledge.[11] Fukuda, we may recall, was the Japanese biologist who had discovered that the prothoracic gland played a key role in metamorphosis of the commercial silkworm, *Bombyx mori.* Williams next confirmed "the great significance of the prothoracic glands" in cecropia when he found that isolated abdomens metamorphosed readily when provided with chilled brains plus two pairs of prothoracic glands.[12]

This was a key realization: the prothoracic glands were essential for metamorphosis, providing the ingredient that Wigglesworth had acknowledged to be missing in his broad overall assessment. Moreover, Williams could now take understanding a vital step further: "From these observations it may be concluded that the brain exerts a controlling action on the prothoracic glands."[13]

In further experiments, published in a series of papers extending into the 1950s, he showed that metamorphosis in cecropia involves three separate steps: (1) it requires the secretion of a hormone by the prothoracic gland, which would fit with Wigglesworth's proposal of a molting hormone; (2) production of this hormone is controlled by a master hormone, then unknown, that was secreted by the brain; (3) but metamorphosis also requires the *absence* of a third hormone secreted by the corpora allata—Wigglesworth's juvenile hormone.[14]

Williams went on to become professor of zoology at Harvard, having by the age of thirty-six established a distinguished reputation in entomology. In 1955, two years after his professorial appointment, Williams traveled to Cambridge, England, on a two-year Guggenheim Fellowship, where he worked in Wigglesworth's lab. The two men did not become close associates, in the memory of others who worked in the lab. Not that there was open hostility between them. Both were too gentlemanly for any unpleasantness, but perhaps professional rivalry and the differences in personality may have proved too great an obstacle.

Wigglesworth, by all accounts, appeared shy and withdrawn, never permitting a free and easy friendship with any colleague. His retiring nature even extended to his children, who would joke that they should have been insects, since only then would he have paid them sufficient attention. He was not devoid of humor, however: in his famous weekly teas, his cackle of a laugh was unmistakable, and a dry humor is noticeable in his more philosophically or publicly directed books and papers.

Williams, in the words of A. M. Pappenheimer Jr., was "a traditional Southerner, a man of charm and unfailing courtesy, who kept his rich Tidewater accent untainted by clipped Yankee tones throughout his life."[15] Simon Maddrell, who was a youthful postgraduate in Wigglesworth's lab at this time, would recall him as a genial extrovert, who walked the Cambridge streets in a ten-gallon hat. Where Wigglesworth worked for the most part in solitary intensity, nearly all of Williams's research papers were collaborations with other workers, often his postgraduates. He appears to have been more extravagantly sociable, with "humorous overtones" even in his scientific papers, a characteristic that emerged as early as his PhD thesis. Indeed, one of his graduate students recalled: "When I think of Carroll's achievements, I am overwhelmed by memories of hilarious events and merry times . . . life in his lab was usually such fun, and we all shared so many laughs."[16] Not everybody came away with such a genial impression, though. In the words of his subsequent research associate, Lynn Riddiford, Williams had a very strong personality and had forceful opinions on the relevance of his work and that of others. "There were people who liked him and people who didn't."[17]

In fact, when one looks a little deeper than the differences in personalities of the two scientists, there were striking commonalities in their attitudes to science. In both we discover a dedication to their work that was unusually intense and pure. This is evident in both the careful planning of experiments and in the massive outpouring of their researches. Only later, when he himself had succeeded to Wigglesworth's chair at Cambridge, did Simon Maddrell come to see that for Wigglesworth (and an American colleague might equally have been talking of Williams), "science was to a large extent his life. He devoted himself to it in a very natural way, not out of duty but driven by a great curiosity about natural things and how they worked, and driven by a great desire to explore this whole new area of insect physiology."[18] This is exactly the determination one might expect to find in a winning Olympic athlete. Where others, perhaps less determined, met with more limited success, the single-mindedness of Wigglesworth and Williams helped them to succeed at a deeper, altogether more profound level.

I was struck, in considering the insect metamorphosis story, by the fact that Wigglesworth, who had pioneered the discovery of juvenile hormone, had never taken his researches to their logical conclusion. Rather he appeared to have abandoned this to Williams. In questioning Maddrell about this, he acknowledged:

"If he had continued his deep involvement with metamorphosis, one would have expected him to move on from *Rhodnius* to the endopterygotes [fully metamorphosing insects]."

I pressed him a little: "He obviously never felt it necessary to do that?"

"No. He took it to the death of what interested him and then he went on to something else."

"Perhaps he just felt that he had taken it far enough when it got down to the more tedious biochemical extrapolations?"

Maddrell shook his head. "I don't think that was the motive. I think I understand what it was, at least partially, now. His attitude would have been: 'There's just some other mystery I'd like to get on to.'"

"I can see that it was a fantastic position to be in—when there was so much that was unknown in the field of insect physiology."

"Absolutely. He created the field."

— 15 —

Ecology's Magic Bullet

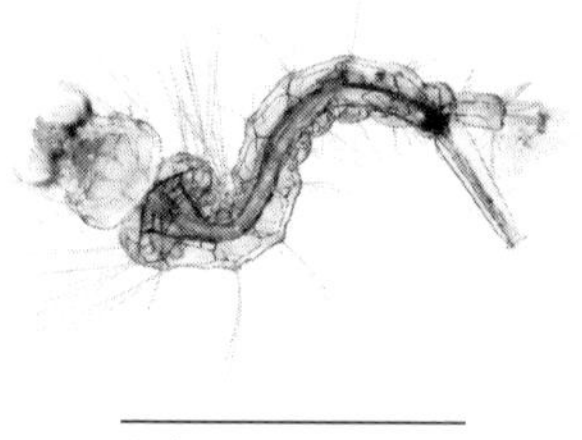

Culex mosquito larva

In 1954, a German biochemist, Peter Karlson, working at the University of Tübingen, together with Adolf F. J. Butenandt of the Max Planck Institute for Biochemistry in Munich, crushed vast numbers of pupae of the silkworm moth, *Bombyx mori*, to accumulate eleven hundred pounds of crude homogenate. From this they isolated less than an ounce of a crystalline extract of purified molting hormone.[1] On analysis, they found that, like the substances secreted by the adrenal glands in humans, it was a steroid hormone. They named it "ecdysone."

At about this time, Carroll Williams was working on ways to extract and purify the juvenile hormone. He began by confirming that excision of the corpora allata from chilled cecropia pupae had no effect on development into normal fertile adult moths—one would expect inactivity of the corpora allata at this pupal stage—but when he tested the corpora allata in the adult moth, he found, to his surprise, that they were more active than at any stage in the moth's life history. Indeed, the highest concentrations of juvenile hormone were to be found in the abdomens of adult male silkworm moths.

During his stay in England Williams made his first crude extracts, using petroleum ether, from pulping large numbers of male abdomens. On his return to Harvard, he produced increasingly pure isolations, work that carried over into the following decade. In 1964, he was joined by Karel Slama, who had been conducting an unrelated line of research in his Prague laboratory rearing the fire bug, *Pyrrhocoris apterus*, from fertilized eggs. This common and widely distributed insect takes its name from its flame-red-against-charcoal coloration, which resembles glowing coals, a resemblance that can be very striking when it forms dense clusters of tens or hundreds of individuals. However, when Slama attempted to replicate this work at Harvard, the fertile eggs failed to mature, just going on to form supernumerary larvae. The difficulty was eventually traced to the paper he had packed into the rearing jars. Only now did the startled entomologists realize that the plant kingdom had beaten science to some useful applications of insect metamorphosis.

When one considers how vulnerable plants are to attack by the multitude of animals that prey on them, one marvels at their endurance. Plants have, of course, evolved physical deterrents of their own, such as thorns, stings, and noxious chemicals. Indeed, as Dame Agatha Christie was well aware, some of the most deadly poisons in the world derive from common plants in the wild, such as deadly nightshade. Some of these poisons help plants to defend themselves against insect predation by interfering with insect metabolism. Williams and Slama had just discovered that natural production of an analogue of juvenile hormone by many different plants was an important ingredient in this protective repertoire. In time, juvenile-hormone analogues would be isolated from a variety of common plants such as the balsam fir, the pepper tree, sweet basil, and sedges. Topical applications of these analogues to final-stage insect larvae were found to block metamorphosis to the adult, resulting in a supernumerary molt to giant immature larvae, much as the entomologists had discovered in their experiments. These juvenile-hormone analogues also found their way from plants into manufactured paper, including the pages of the

New York Times, *Wall Street Journal*, *Boston Globe*, *Science*, and *Scientific American*, which were shown to inhibit the metamorphosis of those same fire bug larvae. When Slama replaced the paper in the rearing jars with the highly purified Whatman filter paper, metamorphosis preceded normally.[2]

Williams was not slow to realize the potential of these findings. Juvenile hormone, or a closely derived analogue, could be useful as an insecticide. Affected larvae would fail to mature, cutting off the reproduction of unwanted pests.

In 1956, he wrote a letter to *Nature* in which he described the discovery of juvenile hormone in American paper. In his letter, he coined the catchphrase, "the Paper Factor," adding that "In addition to the theoretical interest of the juvenile hormone, it seems likely that [juvenile hormone] will prove to be an effective insecticide. This prospect is worthy of attention because insects can scarcely evolve a resistance to their own hormone."[3]

At about this time, Williams had begun working with Lynn Riddiford, who, in an interesting historical footnote, was described as "one of the few women scientists in Harvard history to reach professorial rank."[4] The two scientists were having difficulties in mating the saturniid silk moth, *Antherea polyphemus*, in the laboratory. Realizing that the larvae of *Antherea* feed on the leaves of oak trees, they added a few oak leaves to the cages, whereupon the female appeared to secrete her pheromone, and mating preceded normally.[5]

In a *Boston Sunday Herald* article, in August 1959, Williams was quoted as hoping that the chemical formula for juvenile hormone would be written by the end of the year, which should allow not only its synthesis but also the manufacture of a series of chemical analogues and derivatives. These notions provoked some humorous interludes with the lay public, who equated juvenile hormone with the "elixir of youth," assuming it might act, in Peter Pan fashion, to retard aging in humans. The newspaper quipped: "Crowds of fascinated people cannot keep from guessing what the end results of the Harvard biologist's youth hormone experiments may be." The bemused Williams received torrents of telephone calls, cables, and letters from individuals "seeking to splash in the fountain of youth." He also recounted stories of cosmetic manufacturers who "plead with me to have a little of the stuff in crude form to put into beauty creams and lotions."

"If you could drink the hormone yourself," a reporter asked him, "when would you take it?"

"Right now!" was the tongue-in-cheek reply.

In a series of papers extending into the mid 1960s, Williams and his colleagues went on to demonstrate the production of juvenile hormone by the corpora allata, its role in the endocrine control of molting, pupation, and adult development in the cecropia silkworm, and finally, in 1965, the crowning achievement of its extraction and purification.[6] As both Wigglesworth and Williams had foreseen, this had consequences that extended far beyond entomology, including practical applications that are increasingly relevant to medicine and agriculture even today.

Thanks to these pioneers of the study of insect metamorphosis, we now know that the molting process begins in the "neurosecretory cells" in the insect brain, which send messages along nerves to a pair of tiny organs, the "neurohaemal organs" or "corpora cardiaca," which are located very close to the corpora allata. These in turn release their store of "prothoracicotropic hormone" (PTTH), which is carried by the bloodstream to the prothoracic gland, where it stimulates the production of the molting hormone, ecdysone. The close anatomical proximity of the corpora cardiaca to the corpora allata explains the initial confusion in Wigglesworth's results and in his thinking. To add to the confusion, in butterflies and moths PTTH is actually stored in the corpora allata, so their removal would, in the Lepidoptera, cut off the stimulus to production of ecdysone by the prothoracic glands.[7]

Each molt of the insect larva is triggered by one or more pulses of ecdysone. A first pulse of neuroelectric excitation from the neurosecretory brain cells induces a small rise in ecdysone in the larval blood. This elicits a change in cellular commitment—in other words, it acts as a kind of developmental primer. A second, larger pulse of ecdysone sets the ball rolling for the full changes in tissue and cell differentiation associated with the molt. The location of the priming signal in the brain allows environmental cues to influence the timing of molting, for instance, the abdominal bloating in *Rhodnius* after a blood meal as discovered by Wigglesworth. Another example of control through the same basic mechanisms is seen in the silkworm moth, *Platsamia cecropia*, the subject of Carroll Williams's groundbreaking experiments, in which PTTH secretion abruptly ceases after the pupa is formed. The pupa can thus remain in the suspended state of diapause throughout the winter, after which the temperature rise cues its final molt to the adult. The in-built mechanisms of diapause also explain how *Rhodnius* larvae lived for up to a year after decapitation. An understanding of such processes is important in the silk industry, where the silk-spinning pupa is artificially primed by two weeks' exposure to cold.

Today we also know that Wigglesworth was right in his earliest experiments on the inhibitory effects of the corpus allatum. It does indeed produce juvenile hormone. This blocks metamorphosis in the younger nymphs, larvae, grubs, maggots, and caterpillars of all metamorphosing insects.

Wigglesworth demonstrated this in a very graphical manner when he painted his initials using juvenile hormone onto one of the abdominal segments of a fifth-stage *Rhodnius* nymph before metamorphosis. When metamorphosis was complete, the initials VBW were permanently etched in larval colors and texture on the differently colored and textured abdominal skin of the adult.

Posterity has also confirmed that when the larva is fully grown, the corpus allatum switches off its production of juvenile hormone. This removal of the inhibitory agent results in the metamorphosis to the adult in incompletely metamorphosing insects and in the metamorphoses from larva to pupa, and from pupa to adult, in completely metamorphosing insects, when, under the stimulus of the molting hormone, ecdysone, developmental genetic programs intrinsic to this stage of the larval and pupal tissues effect the changes.

The insect metamorphosis story must rank as one of the most exciting, and instructive, in twentieth-century biology, a majestic demonstration of the power of deductive reasoning and scientific experiment in unraveling a great mystery of nature. For medical epidemiologists working against lethal epidemics such as malaria, trypanosomiasis, yellow fever, and dengue fever, which are spread by insects, such understanding offers the potential for biological control of the diseases.

In 1966, just a year after he and his colleagues had first reported the chemical extraction of juvenile hormone, Williams demonstrated the lethal effects of a synthetic juvenile hormone on the larvae of the yellow fever mosquito, *Aedes egypti.*[8] A year later he wrote a review article for *Scientific American* on the practical applications of juvenile hormone analogues for agriculture as well as human health.[9] That same year he and Riddiford showed the effects of juvenile hormone analogues on the embryonic development of silkworms.[10] In an interview with the *Boston Globe*, the two scientists explained the future ecological potential of such extrapolations of the metamorphosis experiments: "It seems possible to develop hormone insect-killers that will affect only one specific kind of insect pest. Laboratories in several countries are working on this. And the insects probably will not be able to develop the kind of resistance that they build up against insecticides such as DDT."[11] In this, Williams appeared to be predicting the so-called "ecological magic bullet," a perfect treatment much as the nineteenth-century German scientist Paul Ehrlich had predicted chemical cures for infections before the era of antibacterial and antiviral drugs.

Today hundreds of artificially constructed juvenile hormone analogues have been produced. These, and a variety of other extrapolations, have made possible environmentally sensitive pest management through hormonal disruption of insect development, metamorphosis, reproduction, diapause, and other aspects of insect behavior.

One such juvenile-hormone derivative is methoprene, a general-use pesticide that goes under a variety of trade names. Sufficiently environmentally sensitive to be classed by the EPA as practically nontoxic, a mere five grams of methoprene will cover an acre of mosquito floodwater ground. It seems likely that the wide diversity of insect metamorphic and reproductive strategies will, in time, offer an ever increasing range of highly targeted potential solutions. I discovered a remarkable example of this developing biotechnology when I visited Chris Curtis and his colleagues at the London School of Hygiene and Tropical Medicine.

Professor Curtis is an expert in malarial research and control. His research interest is the malaria-carrying *Anopheles* mosquito, which he breeds in vaulted underground bunkers, known as the insectaries, extending out under Gower Street. He has offered himself as live bait while conducting mosquito counts in African villages, sitting out of doors all night with his trouser legs rolled up—a risky dedication to duty that, on one occasion, gave him a bout of falciparum malaria. You could say that he knows his subject intimately. His office wall is adorned with a map of Africa, with colored pins marking the sites of different strains of *Anopheles* eggs that have been ferried to London for breeding experiments.

Malaria has long been a scourge in Sri Lanka, as in many other tropical and subtropical areas of the world. But in 1967, thanks to widespread spraying with DDT, the medical authorities there almost eradicated the disease, with just seventeen cases reported that year for the whole country. All seventeen cases were in a central area, known as Elahera, where there is a tradition of opencast gem mining. Under license from the State Gem Corporation, the miners dig shallow pits by hand in search of gemstones, such as emeralds. They are required, under the regulations, to fill in the pits after they have been exhausted, but in practice many pits are abandoned to fill up with water during the rainy season. These become breeding grounds for *Anopheles* mosquitoes. In the 1990s medical investigators estimated that a nomadic population of 25,000 to 50,000 people were still engaged in gem mining, and the numbers of shallow pits that could form potential breeding habitats for mosquitoes were estimated to be between 247 and 370 per hectare. To make matters worse, the huts the miners occupied were made out of woven palm leaves and polythene, which gave no physical protection from mosquito entry and proved unsuitable surfaces for insecticide spraying.

Traditional preventive measures no longer worked. The result was a resurgence of malaria centered on the gem-mining areas and amounting to hundreds of thousands of cases annually, and a great many deaths. Over the years 1993 to 2000, one of Curtis's postgraduate students, Amara Yapabandara, began working with experts from the Sri Lankan antimalarial campaign. With Curtis's help, they set up a trial of the juvenile hormone analogue, pyriproxyfen, in an attempt to eradicate the *Anopheles* population in the mining area.[12]

As her test area, Yapanandara chose a cluster of eight villages, mapping the area to show the geographical distribution of the houses and any potential breeding places. The location of every house, path, and major water body was surveyed by counting paces along compass bearings. Each house was given a geographic reconnaissance number, which was marked on the door. Within the 1.5 kilometer zone around each village, all the gem pits were also numbered, and mosquito populations in the study areas were estimated based on a variety of sampling methods. Information on malaria cases was gathered from four different field clinics and from the outpatient departments of the local government hospital and dispensary. This involved a huge community operation, with village education and discussion, and periodic mass blood screening in addition to monitoring of infected patients. Of the eight chosen villages, four were assigned to treatment and four assigned to be controls. In the four chosen for treatment, Yapabandara treated every pit, amounting to several hundred per village, together with any other potential sources she had spotted in the initial survey, spraying them with pyriproxyfen twice a year. Mosquitoes were systematically counted and cases of malaria tabulated and the results of the treated villages compared to the controls.

"It worked beautifully," Curtis told me. "There was a dramatic reduction in mosquito counts in the treated villages, and such an effective reduction in malaria that . . . it clearly showed that pyriproxyfen was sufficient to suppress malaria in the treated villages compared with the controls."

As this practical application implies, the metamorphosis research has never actually ended, but continues today, in the field and in many laboratories throughout the world. A lucid summary of the modern perspective on insect hormones is provided by H. Frederik Nijhout, of the Department of Biology at Duke University, who, in a nutshell, declares that "virtually every aspect of the post-embryonic development of insects is controlled by hormones."[13] This includes metamorphosis per se, the timing of developmental processes, the molting cycles of the huge variety of insects, the fates of individuals among the social insects, determining castes of ants, bees, and termites, and the distinctive seasonal forms seen in many species. The hormones responsible for this huge variety are eminently simple: a handful of ecdysones and juvenile hormone are responsible, as Nijhout explains, "for virtually the entire panoply of events." Such is the legacy of the work of Wigglesworth, Williams, and their colleagues from many countries who contributed to the metamorphosis experiments.

But understanding the physiology of metamorphosis in insects, elegant as the hormonal control has proved to be, is not synonymous with understanding the evolutionary origins of metamorphosis, whether in insects, or in its wider extrapolations to the entire animal kingdom. To do this we must cast our net wider. If there are commonalities in the evolution of metamorphosis throughout the animal kingdom, we should bear in mind what we have learned from the study of the physiology in insects, while in the meantime we must look once again to the oceans, where we discover a great many whole phyla, embracing myriad varieties of metamorphosis more baffling and varied than even the strange and beautiful world of the insects.

PART THREE

New Perspectives

Yet to admire only our own successes, as if they had no past (and were sure of the future), would make a caricature of knowledge. For human achievement, and science in particular, is not a museum of finished constructions.

—Jacob Bronowski, *The Ascent of Man*

16

On the Steps of York Minster

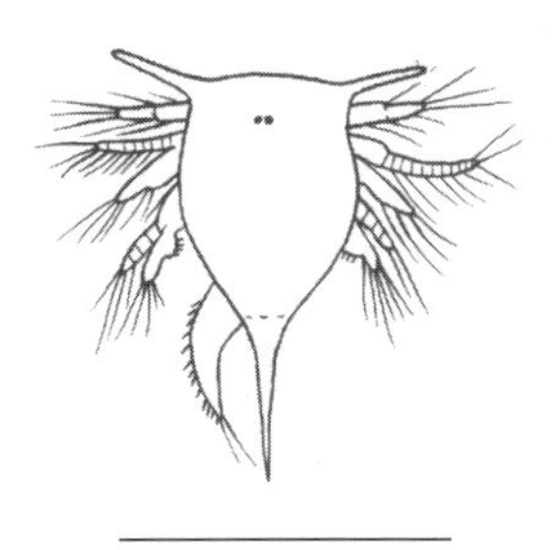

Nauplius larva of barnacle

At nine o'clock in the evening of January 25, 1991, in the freezing cold of the Yorkshire winter, three figures waited on the steps of the cathedral church, York Minster, for the arrival of an American editor named Greg Payne. Don Williamson, wearing padded trousers and huddled under a car rug in a wheelchair, was shivering. He was accompanied by his second wife, Enid, and Mary, his sister, both also shivering despite neck scarves and winter coats. They had arrived early and Payne was late. Mercifully it had not snowed. "We didn't know Greg by sight. We had never met him. People were coming up to speak to us, assuming we were keeping some special vigil or something, though nobody offered us any money." The recovered voice, together with the sense of humor, was quintessentially Williamson. And now, with the prospect of his visitor, the marine biologist had rediscovered his sense of hope and purpose.

After his stroke, Williamson had lost the ability to speak. Fortunately some recovery began when he was still a patient in Noble's Hospital in Douglas. While struggling there to come to terms with his stroke, he had received another rejection of his book.

It had already been rejected by six publishers when Lynn Margulis intervened to encourage two editors she knew to take it seriously. The first of these was an editor-at-large at Yale University Press to whom Williamson sent a copy of the incomplete manuscript. The editor had responded enthusiastically but added that he would need the approval of his editorial board when it next met. Sadly, in the interim, the editor became seriously ill and his successor promptly declined interest in the project.

While this was going on Williamson was facing other difficulties that for many would have proved insurmountable. When he had his stroke he had lost the ability to speak, a condition known as aphasia, which was accompanied by the inability to read. Over time he had made a partial recovery. His aphasia had also been accompanied by dysgraphia, which had impaired his ability to write. When the Yale University Press rejection arrived Williamson initially had difficulty comprehending the editor's letter. When he did manage to understand it, he was devastated by its content.

At this time he shared a two-bedded ward with Robert, a seventeen-year-old youth suffering from recurrent arthritis. Robert was studying for his A-level examinations, which in the UK is a prerequisite for university or college admission. The sixty-eight-year-old scientist and the youth soon struck up a friendship, and Robert became fascinated by Williamson's theory of metamorphosis. But it was an enormous struggle for Williamson to explain it to him, given his difficulties with speech, compounded by Robert's lack of knowledge of biology. Nevertheless, little by little, over the fortnight or so that they shared the room, Williamson explained something of biology, the prevailing evolutionary principles in general, and the challenging nature of his own theory in particular. The exercise proved therapeutic. "Robert asked me intelligent and searching questions, and this made me work through the whole thing and even enlarge on it. It was very hard work but I think it was worth it because it forced me to relearn how to put my thoughts into words."

As Williamson's speech improved, his ability to read and write also began to recover. But he was left with a paralysis affecting his right hand, so he set about teaching himself to type with his left hand. For the normally right-handed scientist, this was made all the more difficult by the fact that he had lost the spatial memory of the typewriter keyboard. He had to reeducate mind and hand, like a child first learning the alphabet, but he persisted, day after day, and finally recovered his typing ability.

All hope for publication of his book now focused on the second of the two editors Margulis had recommended. This was the Gregory W. Payne, of Chapman and Hall in New York, for whom Williamson was waiting on the cathedral steps in York, and to whom Williamson had written on his return home to Port Erin. "In retrospect it seemed rather ironic, because I had already submitted the book to Chapman and Hall in London, who had rejected it, saying, 'When your theory is better known, we will reconsider it.'" Of course, from Williamson's perspective, the whole purpose of publishing the book was to get his theory better known. Once back home, though still affected by his stroke—his recovery would in the long term prove to be a partial one—Williamson switched from a typewriter to an early model Amstrad computer and began, all over again, describing his theory and publication hopes to Payne. To Williamson's joy and relief, Payne wrote back:

"Certainly, we will publish it."

As it happened, Payne was planning to visit Yorkshire, where Williamson and his wife had gone to visit with his sister, Mary. It was Payne who suggested the steps of York Minster as a meeting place. So it was that Williamson, his wife, and sister, were waiting there on that bitterly cold January evening. Payne arrived half an hour late, accompanied by a friend. All five retired to the nearest hotel, where in more congenial surroundings the editor duly produced a publisher's agreement that Williamson signed on the spot. Payne comforted Williamson prior to leaving, "We'll try and have it out for your seventieth birthday."

In the event Don Williamson's book, *Larvae and Evolution: Towards a New Zoology*, was published by Chapman and Hall in

1992, six months after his birthday.[1] He dedicated it to Enid, who had not only supported and sustained him throughout his difficulties but had also drawn many of the illustrations—and there were a great many of these, both wonderful and arcane. Anybody reading through the book for the first time will be struck by its clarity of thought and exposition. Under normal circumstances, it would be considered a laudable achievement. In the real circumstances of its genesis and publication, it was nothing short of remarkable. Lynn Margulis and Fred Tauber wrote the foreword, in which they described Williamson's difficulties in getting his ideas published and made clear their opinion that Williamson's theory deserved the spotlight of scientific scrutiny and the courtesy of a proper hearing.

Williamson's talk in Boston in 1990 had provoked a heated reaction among some of the professional biologists who made up his audience. A book is a more comprehensive vehicle than a lecture, allowing more detailed and comprehensive argument. In the preface, Williamson attempted to mollify the inevitable criticism by explaining that he was perpetrating only a minor heresy. By and large he agreed with, and admired, Darwin. He accepted Darwin's idea of evolution as the explanation for the diversity of life, and he believed that natural selection had played a fundamental part in that process. What he disagreed with, in essence, was the universal insistence on the classical branching-tree depiction of what is known as "speciation." Darwinians believe that the formation of new species comes about through branching from an existing twig on the tree of life—in essence a linear process of divergence. But Williamson contended that the only logical explanation for what he had observed in the evolution of some animals and their larvae was that occasionally some branches had actually met and fused at some point during their evolutionary history.

This reunion of branches, as opposed to the linear process of branching, constituted a "reticulate" pattern of evolution. Moreover, in Williamson's opinion, it was not confined to neighboring branches, such as species within a genus, but extended much wider to include the fusion of very different evolutionary groups. This resulted in the coming together of pre-evolved whole genomes—the sum of all their genes—of different life-forms. To Williamson's way of thinking, this did not refute Darwin's theory: it might best be interpreted as a modification of Darwinism as most biologists currently interpreted it. His book went on to explore the links between the developments of animals, their evolutionary history, and many examples where hybridization might have resulted in reticulate fusions of different evolutionary pathways, to give rise to some of the present-day life histories of metamorphosis.

He began with the crustaceans, explaining the anomalies he had observed before moving on do the same for the entire phylum of the echinoderms. The deeper he looked at this colorful and varied group, including sea urchins and starfish, the stranger the life histories, and by implication, the evolutionary origins, of these familiar creatures appeared to be.

We have already seen that all living adult echinoderms are radially symmetrical in the horizontal plane. In the vertical plane they do have an "oral" underside, where you find the mouth, and an "aboral" top side, where, in most echinoderms, you find the anus—but there is no back or front, no left or right. The outer surface is armed with the calcareous spines typical of their phylum. The underside also bears soft, flexible tube feet, which enable the animals to creep over the ocean floor. The tube feet are connected to an internal system of fluid-filled canals, the water vascular system, which can extend and contract the feet. In many species the tube feet are also fitted with suckers to enable them to attach to the ocean floor and also to capture prey.

The echinoderms have a tube-shaped gut that runs through an internal cavity, the coelom, into which the gonads protrude. The coelomic fluid is circulated internally by cilia. With only rare exceptions of internal fertilization, eggs and sperm are shed through this same ciliary action into the sea, where they are fertilized and left to survive on their own. A single starfish can extrude as many as two million eggs in one spawning. Startling as this might seem, it is no more prodigious than the numbers of sperm ejected in a single male emission in humans.

A striking and little realized similarity is also found in women, where the ovaries protrude into the human version of the coelomic cavity, and the eggs are also shed into it, whence they make their way into the collecting tentacles of the fallopian tubes, then pass along the tubes and down into the uterus, which has an open connection, through the vagina, to the environment. Here we discover a reminder of our distant oceanic past, for when women bathe in the ocean, their coelomic cavity is connected to the ocean, however tenuously, and in such circumstances human eggs, if not fertilized en route to implant in the uterus, would also be shed into the ocean.

Echinoderms possess a simple heart, which circulates their equivalent of blood through a system of hemal channels. But other aspects of the echinoderm's inner structure are altogether strange when compared to our bilaterian inheritance. Echinoderms have nothing resembling a brain. Instead, nerves radiate from a ringlike structure around the mouth and branch off to all parts in contact with the outer environment. All the same, this is enough for these intriguing creatures to control and coordinate the necessities of daily life, sensing prey, making purposive movements toward it, and consuming their food and excreting their waste.

There are considerable variations on this basic theme among the different classes of echinoderms. Where starfish creep on their tube feet, equipped with suckers, brittle stars use their long snakelike arms for locomotion. Curiouser still, as we saw with Kirk's brittle star, where starfish have a gut serviced by a mouth at one end and an anus at the other, brittle stars have a single opening to and from their digestive tract through which they both ingest their food and void the waste contents of their gut.

Sea urchins, heart urchins, and sand dollars have no radially distributed arms. Instead the body is enclosed within a rigid shell of closely fitting plates that support movable spines and openings through which the tube feet protrude. Like their starfish cousins, sea urchins have both a mouth and an anus. Their coelom-placed gonads are considered a delicacy in certain countries, where they are farmed for food. Sea urchins and sand dollars are equipped with five teeth, which enable them to graze food from hard surfaces. Heart urchins are toothless and so are not capable of grazing. Instead they live in tubes in the sand where they move particles to their mouths using modified tube feet.

Sea cucumbers are different again, with member species shaped somewhat like the vegetable that gives them their name. Sea cucumbers are also consumed as food in certain parts of the world—though they taste rather different from the vegetable cucumber. Feather stars and sea lilies, respectively, resemble a beautiful mass of colorful feathers and look like clusters of long-stemmed flowers. Though both feather stars and sea lilies are animals, they have phases in their life cycle when they are tethered to the bottom or to driftwood and are among the most gorgeously decorative residents of coral reefs. Finally there is a strange, newly discovered class of echinoderms, the sea daisies, which resemble their namesake's flowers, with tiny heads a centimeter or so in diameter. These are known from three species within the genus *Xyloplax*, which were discovered in sunken, waterlogged wood in deep water off the coast of New Zealand.

In his book Williamson considered the puzzling aspects of all of these diverse classes of echinoderms, with their differing life cycles, explaining how the enigmas might be interpreted in terms of his larval transfer theory, before bringing the whole to a cohesive evolutionary synthesis.

Darwin adopted a linear theory of evolutionary progression, with natural selection operating, through small steps at a time, over vast time periods. This incremental and linear process operated at every stage of the individual organism, from fertilized egg to embryo, and then right through to the adult stage. Today, most evolutionary biologists still espouse this view, which allows for evolutionary innovation to take place at any stage of the life cycle. But Williamson did not subscribe to any all-embracing interpretation of Darwinian linearity. "The theories I am now putting forward imply that this assumption is frequently invalid," he noted in his book.

He proposed that the larval body plan sometimes derived from an entirely different evolutionary lineage from the adult body plan. If true, this had major implications for the evolutionary history and genetic inheritance of marine life-forms. Evolutionary biologists from Darwin onward had looked at the structures of embryos and larvae as important pointers to the evolutionary history, and place on the tree of life, of the adults. Darwin himself had pioneered this approach in his famous examination of barnacles.

As we have seen, the mollusks include a wide diversity of forms and lifestyles, from the familiar shellfish of beaches and seashores, such as mussels and whelks, to the seagoing octopuses and jet-propelled nautili. Most mollusks have a soft body, surmounting a muscular foot, and they are protected by the secretion of an external shell. On superficial examination, barnacles appear to fit this bill. But when Darwin studied their metamorphosis, he made an important discovery. Mollusks typically hatch from their eggs as a trochophore larva—the wheel-like ciliated larvae we saw in chapter 6—but Darwin discovered that barnacles metamorphose through a nauplius larva, pictured at the opening of this chapter, and which are only found in a single phylum—the crustaceans, such as shrimps, crabs, and lobsters. From this Darwin correctly deduced that barnacles were not mollusks at all but crustaceans, as "a glance at the larvae shows this to be the case in an unmistakable manner."

But now, in his book, when Williamson looked in some detail at the life cycles of the crustaceans, the extrapolation of the larval types to their corresponding adults did not always work true. Indeed, as Williamson investigated this further, the extrapolation to the tree of life derived from the larval forms could be very misleading.

He looked in more detail at the specifics of evolutionary development and its implications for metamorphosis.

Some features of the metamorphosis of the echinoderms are much the same, regardless of the type of larva. Paired abdominal (coelomic) sacs form inside the body. One of these, usually on the left, grows bigger than the others, and it is here that the next step in the echinoderm's metamorphosis begins. In the wall of the coelomic sac, pluripotent cells—the equivalent of stem cells—grow into the juvenile adult. This organism develops five lobes, which become radially arranged. From this strange beginning, the form of the juvenile adult echinoderm appears. In many cases the embryonic development really does amount to a new beginning, a novel organism developing according to a new blueprint to construct the adult animal. In echinoderms the pentaradial symmetry owes nothing to the larva. And the bulk of the adult structures, including the arms of the adult starfish or brittle star, develop independently of any larval arm or lobe. Indeed, the adult starfish and brittle star appear to originate and grow within the larva like a separate and independent parasitic being.

Most astonishing of all, in the development of both starfish and sea urchin, the mouth of the adult arises as an entirely new structure unrelated to the mouth of the larva. When one considers the importance of the evolution of the mouth in developmental biology, its position and role determining the great divisions of protostomy, where the first dimple, or blastopore, becomes the mouth, and deuterostomy, where the blastopore becomes the anus, it really is difficult to imagine how, or why, such a vital structure could be abandoned as part of a linear descent-with-modification trajectory. In particular, Williamson highlighted the fact that, although the larvae are clearly deuterostome in their developmental patterns, the term is actually meaningless when applied to the adults—since the mouth of the adult is no longer derived from the blastopore of the embryo.

We have seen how the role of the special pluripotent cells lining the coelomic sac is similar to the role of the imaginal disks during the catastrophic metamorphic change and reprogramming in the insect pupa. So striking is this commonality, it makes one wonder if there could be some evolutionary or developmental commonality.

In both insect and marine species that metamorphose through the activation of stem cells, with the accompanying cataclysmic changes, there appeared to be two separate developmental blueprints involved. For most biologists this was explained through the separate phases of evolutionary adaptation for each phase of metamorphosis within the same organism. But for Williamson, the more likely explanation lay with the amalgamation, through hybridization, of two separate evolutionary lineages.

In chapter after chapter, Williamson outlined his reasons for proposing that the entire echinoderm phylum had evolved from directly developing ancestors that were radially symmetrical throughout their lives. Then he went on to highlight similar anomalies between larval and adult evolution in many other phyla, including the familiar mollusks. In these various marine phyla the adults were radically different from each other in form and life cycles, yet some members of each of the phyla hatched as wheel-like trochophore larvae. The trochophore is as unique in its appearance and structure as the easel-shaped pluteus of the echinoderms. Because of their common trochophore larvae, these different phyla, with their radically different adult animals, are assumed to be offshoots of the same branch of the evolutionary tree. But Williamson questioned what evolutionary commonality the bottom-crawling multi-bristled polychaete worm, *Nereis*, for example, could possibly share with the abalone mollusks, decked out in their beautiful spiral shells and crawling over the rock bottoms on a single powerful foot? In his view, a close evolutionary link between these very different phyla was unlikely.

In Williamson's book there are illustrations of metamorphic development from these common trochophores that defy simple verbal description. For example, the metamorphosis of the polychaete worm, *Owenia fusiformis*, is more spectacularly bizarre than Ridley Scott's *Alien*. As Williamson explains, "Few of those who have favored convergence as the explanation of the similarities have given any evidence for their views." It would, indeed, be hard to do so since the huge evolutionary differences between the phyla were already apparent at the beginning of the Cambrian period, some 570 million years ago. Moreover, unlike the very different adult developments, the development from fertilized egg to trochophore larvae in these different phyla is very similar, suggesting that the common larval shapes really do derive from a common evolutionary origin and not convergence. He challenged traditional evolutionary biologists to explain the minimal changes between larvae compared to the remarkable differences between the adult forms in the same animal whose life cycle is purported to have existed for hundreds of millions of years.

Whatever one's view of Williamson's theory, it's clear that he possesses formidable knowledge of marine invertebrate larvae, and he makes a good argument. There would appear to be glaring anomalies in the evolutionary relationships of marine larvae and adults. Yet Williamson himself had to admit that there were many questions that remained to be answered in his own theory. How could he distinguish a larval form that had, in his opinion, been transferred from one taxon to another from one that had evolved in a gradualist and linear manner? Did the transfer, if and when it happened, result from an isolated hybridization between individuals of different species, or did it represent repeated crosses between different groups? These questions were not easily answered. Moreover, there was a significant omission that ran through all of his theorizing. He had no knowledge of what was happening at the genetic level. More than anything, this lack of a genetic input had damaged his credibility in the eyes of the American colleagues listening to his lectures. Another difficulty lay in the fact that, in 1992, hybridization languished as an unfashionable subject for budding evolutionary biologists. The genetic implications of the union of two different genomes, with all of their complex dynamics and infrastructure, were daunting. Yet hybrid experiments appeared to work—and hybrid organisms certainly existed. How would a hybrid history affect the developmental pathways of an organism? Nobody knew.

Without resolution of the genetics question, the potential implications of hybridization as an evolutionary force would remain unknown, and this, in turn, would continue to undermine Williamson's credibility with his fellow scientists. Then, out of the blue, in February 1994 Williamson received a letter from Michael

Hart, a postdoctoral fellow working at the Institute of Molecular Biology and Biochemistry at the Simon Fraser University, Burnaby, British Columbia. Hart enquired about Williamson's 1990 sea urchin experiment. He informed Williamson that he was a geneticist by training and was offering to look at the genetics of the hybrid offspring.

17

The First Genetic Testing

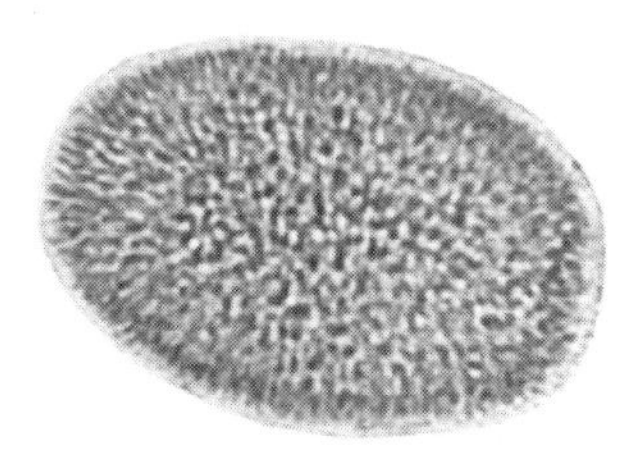

Planula larva of jellyfish

Michael Hart had read Williamson's book in 1992 when he was a graduate student working with Richard Strathmann, a zoology professor at the University of Washington. Strathmann is a distinguished marine biologist, director of the Friday Harbor Laboratories, located on San Juan Island, off the Pacific coast of Washington State, and his stated research interest is "why the beautiful and diverse patterns of development have evolved as they have instead of in other ways."[1] Hart and Strathmann found Williamson's *Larvae and Evolution* challenging, though they remained extremely skeptical about its content. Strathmann had, moreover, written a thoughtful if critical review of the book for the *Quarterly Review of Biology*.[2] Even today the review reads as determinedly open-minded while encapsulating the skepticism of many of his colleagues to Williamson's iconoclastic ideas. In the opening paragraph, he comments, "If the hypotheses and experimental results in this book are correct, then much of evolutionary theory, systematics, and developmental biology will need revision." Williamson, in Strathmann's opinion, only needed to achieve plausibility, not concrete proof, to cast doubt on some of evolutionary biology's fundamental assumptions. To achieve plausibility, Strathmann went on, Williamson's theory needed to provide convincing answers to three separate questions.

First, did the larval-adult incongruities Williamson highlighted merit his claim for a radically different evolutionary process, such as hybridization, or could they be accounted for by orthodox explanation? In Strathmann's opinion, convergent evolution could produce incongruities between larval and adult developments. Even the cataclysmic changes seen in echinoderm metamorphosis could be seen as an adaptation for rapid conversion from one developmental form and habitat to another. Second, were the developmental consequences of Williamson's purported hybridizations consistent with the prevailing evidence of what hybridization does to normal developmental processes? Strathmann did not think so. But he acknowledged that hybridization and its genomic implications was specialist territory and would require a developmental biologist to present a more detailed critique. At the end of his book, Williamson proposed a number of biochemical, immunogical, and genetic tests that would, in his opinion, confirm that his hybrids were genuine. These were the basis of Strathmann's third line of questioning. Were the proposed tests going to be sufficient? Strathmann doubted this. Williamson had not focused on genes important for structural development, so none of these tests would be capable of confirming his more general hypothesis that traits of adults and larvae from different taxonomic groups had been combined into one life history by transfer of genetic material.

Strathmann looked at Williamson's hypothesis as tantamount to believing in miracles, and he quoted the philosopher David Hume's line: "Miracles are by definition highly improbable events." Yet in spite of his doubts, and in spite of what he judged to be the manifold weaknesses of Williamson's hypothesis, Strathmann believed that publication of Williamson's book had been appropriate: "If a revolutionary hypothesis is suppressed, it cannot be evaluated." After all, genetic transfers across the species barrier were known to occur, for example, through symbiotic merging of distantly related life-forms. And even small-scale genetic changes might have major effects on body growth and form. He perceived an important general implication of Williamson's experiments: "Could transfers of genetic material between distantly related organisms have important morphogenetic [developmental] consequences?" He could not offer an example of such a scenario, "but attempts to construct one could succeed or fail instructively, and much hangs on the answer."

Where so many biologists and editors had merely dismissed Williamson, or worse, attempted to prevent him from publishing his theory, a more open mind saw beyond his experiments, and even beyond his larval transfer theory, to an opportunity for greater scientific enlightenment. In an important sense, and this is the sense in which Strathmann saw it, Williamson's ideas and experiments would advance the cause of science whether they stood or they failed. But what more did Williamson have to do to convince the world of biology to take him seriously? Strathmann was in no doubt: "Positive results from hybridizations of distantly related animals would directly confirm Williamson's proposed basis for adult-larval incongruities."

It is to Strathmann's credit that he attempted to repeat Williamson's experiment with species locally available in the San Juan Archipelago. However skeptically he referred to his own cross-phyletic hybridization experiments as attempts at "cold fusion," he nevertheless encouraged other biologists to do the same. "Why not," he urged, "when it is so easy?" This is exactly what Williamson had been asking his colleagues to do for years. Strathmann had some of his students try to cross tunicates with echinoderms, Hart later told me, but these experiments failed to produce embryos. Hart also confessed that he was just as skeptical as Strathmann when he read Williamson's book in 1992, "but we both thought it was worth pursuing or at least thinking about."[3] This had led him to contact Williamson in search of viable material from the 1990 hybridization experiment.

We might recall that in this experiment Williamson had crossed the eggs of the sea squirt, *Ascidia mentula*, with sperm of the edible sea urchin, *Echinus esculentus*. Out of twenty-five attempts at cross-phlyletic hybridization, the great majority had produced no larvae at all. In six attempts, eggs had hatched as free-swimming forms that resembled "ciliated blastulae." In three experiments, larvae had developed beyond the gastrula stage, producing the paternally derived pluteus larvae in various numbers. About 75 percent of the hybrid plutei were stunted or showed some deformity, for example, an arm failing to develop. Similar stunting or deformity of larvae had been witnessed by other biologists conducting hybridization experiments. About 25 percent of the presumed hybrid larvae showed no deformities and seemed indistinguishable from normal sea urchin larvae. But none of these metamorphosed. Williamson repeated the experiment, this time producing larvae that did metamorphose. Indeed, the new experiment produced over 3,000 pluteus larvae, which Williamson cultured further, supplying them with diatoms as food. More than 70 of these larvae developed juvenile rudiments (the first stage of adult development) inside their coelomic sacs, and 20 successfully metamorphosed over the subsequent 37 to 50 days. These were placed in a small aquarium tank, fed, and monitored. A year after first fertilization, 4 of these survived as apparently healthy sea urchins. Williamson hoped that these would go on to maturity so he could breed from them, crossing one hybrid urchin with another (if he obtained different sexes) and crossing hybrid urchins with non-hybrid urchins and also with ascidians.

Of the pluteus larvae that had resulted from this last experiment, several hundred attained full larval development, but without producing any trace of juvenile development inside them. These continued to swim actively for a time before gradually resorbing their pluteal arms and condensing into a spherical ball of cells with a large rounded protuberance at one side. From this protuberance, these "spheroids" could fix themselves to the walls of the rearing bowl with threadlike organs of attachment. They were seen to release themselves, swim away, and reattach in similar fashion elsewhere. In Williamson's words, "Although they developed from pluteus larvae, they are quite unlike any known stage in the development of an echinoderm." He transferred the spheroids to a small aquarium at 59°F (15°C) and waited to see whether they developed further: however, they all died over time without further evidence of development.

Commenting on this experiment in his book, Williamson freely admitted that there was no evidence that sea squirts and sea urchins had ever hybridized in nature. Crossing them experimentally did not prove that an ancestor of any animal group ever acquired a larval form by cross-fertilization.[4] But it convinced him that he was at least on the right track: "The experiment shows, however, that the first step in the process of transferring a larval form from one species to a very distantly related one by cross-fertilization is feasible."

Hart was skeptical of such cross-fertilization because of his knowledge of developmental biology. To his thinking, Williamson's description of hybrid eggs from an ascidian mother developing into the paternal (echinoderm) pluteus larvae suggested that there was a wholesale takeover of the maternal developmental program by the paternal genome. "The reason I found his example involving tunicate eggs and sea urchin sperm so remarkable—and unexpected, and needing corroboration from other evidence—is that tunicate eggs and embryos have a highly deterministic developmental program. Early [division of the egg], movement of cells, and the fate of these cells, happen without the [fertilized egg's] expression of either the maternal or paternal part of [its] genome. It is all run by maternal messenger RNA plus interactions among the cells of the early embryo. So it is not clear how a sea urchin sperm nucleus could provoke a tunicate egg to become a sea urchin embryo and larva."[5]

It was unfortunate that when Hart wrote to Williamson, in 1994, the plutei and spheroids had all died off over the four years since the hybridization. These had been by far the commonest larval offspring and would have been the ideal source for genetic investigation. Williamson had plenty of material preserved in formalin, but at the time he and his colleagues were unaware that, had they preserved tissues in alcohol instead, they might have been suitable for DNA analysis. However, three of the urchins that had developed from the paternal pattern of pluteus larvae were still alive. He and his wife were about to take a holiday, visiting his sister again in Yorkshire.

He tweaked tube feet from the living hybrid urchins, cooled them in a fridge, put them in a thermos flask, and took them with him. Tube feet will remain viable in these circumstances for at least a week. From Thirsk, he traveled twenty miles south to visit Tom Crompton, a research student at York University. Crompton froze the specimens at –94°F (–70°C) and sent them in an insulated container to Hart, in Washington State.

On receiving the samples, Hart formulated his experimental method to examine their DNA content. There are two specific sites in any animal that contain distinctively different DNA. The great majority of DNA is found in the nuclear chromosomes. This is the common inheritance of the nuclear genes from both the maternal and paternal line. The other site is in mitochondria—tiny power-producing packets in the part of the cell outside the nucleus. Mitochondria were once free-living, oxygen-breathing bacteria that were symbiotically incorporated into the nucleated cells of the forerunners of all plants and animals, and they still contain a residuum of bacterial-type DNA. Mitochondria are inherited solely through the eggs of the maternal line in mammals—but the situation in sea urchins appears to be more complex, with some reports of sperm contribution to the mitochondrial DNA of the offspring. In Williamson's experiment, as noted, the maternal line was the sea squirt *Ascidia mentula*, and the paternal line was the sea urchin *Echinus esculentus*. Hart set out to investigate the genetics in two ways. He used the polymerase chain reaction (PCR) to amplify a portion of one of the mitochondrial genes, which should be maternally inherited, and he compared the nucleotide coding of this fragment with known DNA sequences of the same gene in ascidians and sea urchins. This gene was known to be highly conserved. He next used PCR to amplify a portion of the nuclear 28s ribosomal RNA gene of both species, again a gene known to be highly conserved.

The mitochondrial DNA sequences of the putative hybrids were almost identical to those of the sea urchin *Echinus esculentus* but showed 63 nucleotide differences with the tunicate *Ascidia mentula*. The nuclear gene comparisons were technically more difficult to compare, but Hart was also convinced this fitted with the known sequences of sea urchins. Overall he could find no sequences typical of the sea squirt. Hart concluded: "These results indicate that the putative hybrid developed from a sea urchin egg fertilized by a sea urchin sperm, and not from a tunicate egg as described by Williamson."

These genetic results led to a difference of opinion between the two biologists. Hart did not suggest that Williamson had attempted any deception. Williamson had freely provided him with the tissue samples and had encouraged the testing of his hypothesis. Even though the results had contradicted his expectations, Hart still commended Williamson's willingness to "hazard an unpopular idea" and "his good humor in response to skepticism." Indeed, Hart now produced a theory of his own to explain Williamson's findings. Some sea urchins were known to be hermaphrodites. They produced both eggs and sperm and were capable of self-fertilizing their own eggs. If the sperm collected from the sea urchins had been contaminated with self-fertilized eggs of this nature, it would explain those experimental offspring that had gone on to form pluteus larvae and metamorphose to sea urchins. Williamson refuted this hermaphrodite theory. Given the large numbers of plutei larvae, Hart's hermaphrodite hypothesis would suggest that most if not all of the eggs used were sea urchin rather than tunicate, an inference that the experienced marine biologist rejected. He argued that the eggs of the two different phyla look distinctly different, with sea urchin eggs surrounded by a layer of jelly, whereas ascidian eggs are surrounded by "follicle" cells. He had taken considerable care to exclude hermaphroditic sea urchin eggs as part of his experimental method. Moreover, Dr. Hilary B. Moore had studied the population of *Echinus esculentus* on the Port Erin breakwater in the 1930s, and he found that only 1 in 3,000 was hermaphrodite, making hermaphrodism very unlikely as the explanation of his experimental findings.

Perhaps critical to this difference in opinion between Hart and Williamson was the fact that the urchins that provided the material for gene analysis came from only a small minority of the pluteus larvae. Over 90 percent of the plutei in the same culture had retracted their arms to become spheroids, an unknown form in the development of sea urchins. The great majority of the hybrid offspring, which had hatched to the paternal pattern of plutei larvae before developing to spheroids, had not been genetically examined.

Despite Williamson's protests, Hart went on to publish the results of his genetic examination under his name alone, in 1996.[6] In the publication he described Williamson's idea as an "improbable hypothesis [that] implies that the evolutionary history of the major [evolutionary groups] is a . . . web of hybridizations and divergences." In describing why he had undertaken the experiment, he wrote: "Williamson's heretical view . . . was widely dismissed, though some have advocated testing his experimental results." In 2004, I wrote to Hart and asked him if such language suggested a bias against Williamson's theory at the outset. He replied promptly, and in some detail. He had conducted the genetic examination while an unsecured postdoctoral fellow on a two-year appointment and looking to secure a permanent academic appointment. He had not been biased against Williamson or his hypothesis. Indeed, genetic analysis of the hybrids had seemed a wonderful opportunity. "It would have been an enormous boost to my career (not to mention Don's hypothesis) if I had found tunicate DNA in his hybrid sample. Far from being disposed against his ideas, I was in fact rather disappointed at the bland predictability of my results."

Whatever their differences, both biologists are undoubtedly sincere and dedicated in their respective approaches to scientific truth. But Hart's conclusions were a crushing blow to Williamson's hopes of genetic confirmation of his results. In the introduction to his paper, Hart wrote, "Testing improbable or heretical hypotheses ensures that such hypotheses succeed or fail on their merits and not on their relation to orthodoxy." The implication, as far as Hart was concerned, was clear. Williamson's hypothesis had failed to stand up to rigorous investigation and should be dismissed.

— 18 —

The Evolutionary Potential of Hybridization

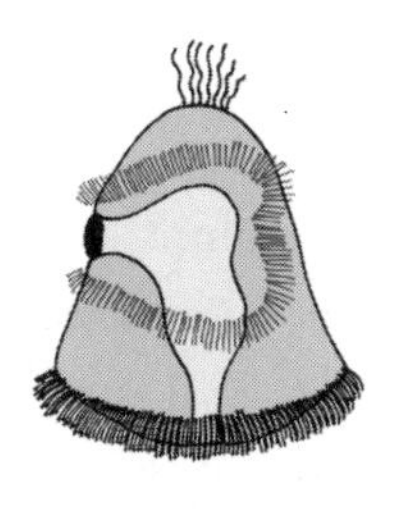

Hypothetical dipleurula larva

Rudolph A. Raff is professor of biology at Indiana University and director of the Institute for Molecular and Cellular Biology there. He has long been interested in the metamorphosis of sea urchins. Hybrids of different species, genera, and even families of sea urchins are known to develop well, undergoing successful metamorphosis. Raff has been particularly fascinated with the elimination of the larval phase as part of the evolutionary development in certain species of marine invertebrates, this being one of the two alternative versions of direct development, as opposed to indirect development through a larval stage. In a 1987 paper, Raff observed that direct development has evolved independently several times, and in several different ways.[1] For him this offered a window onto the study of evolutionary modification. Over decades, Raff and his colleagues have studied the effects of hybridization between two different species within the same genus of sea urchins.

The purple sea urchin, *Heliocidaris erythrogramma*, is the commonest found around the south Australian coast, ranging from Port Stephens in New South Wales to Shark Bay in Western

Australia, where it carves out hemispherical hollows in the rocks at the lowest tide levels. Colors can vary dramatically, from a light olive green, through pink to dark purple, causing a previous generation of biologists to mistake the purple sea urchin for three separate species.

The brown sea urchin, *Heliocidaris tuberculata*, is also a common Australian resident, found in a wide distribution from southern Queensland to mid–New South Wales. Its main distinguishing features are the bright orange color of its spines and, in contrast with the purple sea urchin, the bluntness of its spines at the tip. The brown sea urchin develops through the normal metamorphic phase of a pluteus larva, which feeds and grows in the surface waters for six weeks or so before metamorphosing to a juvenile adult; the purple sea urchin, in contrast, undergoes a simplified pattern of direct development, which includes a giant—relatively speaking—potato-shaped larval phase, devoid of mouth, gut, arms, or lobes, before metamorphosing into the juvenile after just four days.

Studies have suggested that the two species diverged from a common ancestor about ten million years ago, and some time thereafter the purple sea urchin larva lost its feeding structures, even its gut. The purple sea urchin's life cycle has gained many new features in the meantime, including changes in embryo development, notably in the early cleavage pattern of its ball of cells and a novel mechanism of gastrula formation—the stage where the embryo has developed into a hollow ball, with a primitive gut.[2] This has been accompanied by a major increase in yolk, essential to the purple sea urchin larva's non-feeding life history. There have been so many other complex adaptations of the two species, including evolutionary modification of the animal's sperm,[3] that if an observer ignorant of the adult developments were to compare and contrast the present-day pluteus larva of the brown sea urchin with the fattened, yolk-rich, and relatively featureless larva of the purple sea urchin, they would conclude that they were phases in the life cycle of different life-forms.

Raff and his colleagues have developed considerable expertise in hybridizing the two species of urchins, and, in a landmark series of experiments, they have studied the developmental outcome. Hybridization has proved remarkably successful when the purple sea urchin provides the eggs. Reverse crosses, involving the eggs of the brown sea urchin fertilized by purple sea urchin sperm, had to be abandoned after they were found to be nonviable. Using state-of-the-art genetic and molecular biology assessment, Raff and his colleagues have worked out some of the complex accommodations that occur at genetic and molecular levels within the hybrids. It has proved to be an enlightening enterprise.

Past prejudices dismissed hybrid animal offspring as inferior to those of "pure" or "naturally breeding" parents. Sometimes this is the case—but it is by no means an absolute rule. For example, when biologists investigated the outcome of hybrid crosses between two different species of sea urchins within the *Echinometra* genus in the coastal waters of Okinawa, they found that fertilization was high, and the resulting hybrids were vigorous in their growth and metamorphosed to fully reproductive adults.[4] However, the investigating scientists found little evidence for frequent hybridization in their oceanic habitat, although they could offer no obvious explanation for this, since the two species shared the same ecology and their breeding times overlapped. Elsewhere the same investigators found abundant evidence of hybridization between other species of sea urchins in their natural habitat.[5]

When Williamson first presented his larval transfer theory, the great majority of evolutionary biologists were skeptical about any role for hybridization in animal evolution, assuming that hybridization led to genetic instability and infertile offspring. But biologists today are becoming increasingly aware of the importance of hybridization in certain instances of animal evolution. How interesting, therefore, that the various hybrids born of these interspecies crosses are often described as "robust," indicating that the two parental genomes, whatever their evolved differences, appear to have united in a seemingly healthy way in the hybrid genome. Closer study of such hybrid offspring has revealed many surprises.

Raff knows that hybrid "echinoids" tend to follow maternal

features up to and including the formation of the gastrula, after which they tend to switch to paternal features. This observation may have relevance to Williamson's findings of paternal plutei patterns of larvae in his sea urchin–sea squirt crosses, since it suggests that genomes may contain inbuilt organizational switches that trigger different developmental roles at different stages of the process of development. It was interesting, therefore, that as Raff's hybrids switched from a maternal to paternal pattern of development, the purple and brown sea urchin hybrids developed feeding pluteus structures, which had been lost to the maternal line for millions of years. Indeed what Raff and his colleagues observed was far more complex than merely a reversion to the paternal pluteus. The hybrid larvae also developed along lines that were distinctly different from either parental species.[6]

Initial cleavage of the embryo followed the maternal purple sea urchin pattern up to and including the formation of the gastrula. Under normal circumstances, and after the gastrula is formed, a larva following the maternal developmental pattern would have adopted the relatively featureless shape of the purple sea urchin larva, devoid of arms or a gut and with only a single, incomplete, ciliary band. What actually happened is that, over days two to three, the hybrid larva changed to a deeply lobed, flattened form with distinct mouth and anus separated by a single highly convoluted ciliary band. Raff suggests that this might be a most unusual larval development.

We might recall the protostome versus deuterostome branching close to the basal trunk of the conventional tree of life, where the first opening in the developing embryo—the blastopore—becomes either the future mouth, as in the protostomes, or the future anus, as in the deuterostomes. It has long been held that there are two basic types of "primary" or "original" invertebrate larvae, corresponding to those same basal branches of the tree of life. The primary larva of the protostomes is assumed to be the familiar trochophore, the wheel-like larva of many species of mollusks, meanwhile the primary larva of the deuterostomes is presumed to have been modified and changed over the vastness of evolutionary time. But developmental biologists had formulated a hypothetical primary larva of the deuterostomes, based on the deuterostome embryonic development and on commonalities among the various deuterostome larvae—a hitherto unseen larval form they called the "dipleurula." In the words of developmental biologist Pat Willmer: "the dipleurula does not strictly exist as a larval form, and may never have done; it is merely a [theoretical] archetype."[7] Now, it would appear, if Raff and colleagues are correct in their extrapolation, that they had, however fleetingly, resurrected the archetype as a developmental phase in a real organism.

Moreover, as the hybrid larva developed and changed, it exhibited a range of "ancestral" features, some of which resembled the bipinnaria larva of the starfish and the auricularia larva of the sea cucumber. They observed how the hybrid larva continued to change, until, by day five, it resembled the paternal pluteus. But Raff and his colleagues took pains to point out that this later-stage hybrid larva did not represent a reversion to a pluteus. Indeed, in many important respects the hybrid larva represented neither a paternal nor a maternal larva. "We conclude that the developmental mode is truly a 'hybrid' one, in which features from both parents combine to generate a new pathway."

After seven days the hybrid larvae metamorphosed to juvenile adults. Although the development of the juvenile was often imperfect, the biologists concluded that the hybrids revealed "an amazing and unexpected potential for integration of the disparate parental pathways," shedding new light on ways in which embryonic and larval development can be radically reorganized by hybrid crosses while at the same time allowing a coherent development. Unfortunately the resulting young adults did not live long enough for the biologists to be able to perform cross-hybrid crosses or backcrosses with either of the parental species.[8]

To Raff and his coauthors the unique structural and genetic pathways revealed by the hybrids suggested that "[this] might be a way in which developmental novelties can arise." And although they were too cautious to say so, it also suggests that major evolutionary change might come about through hybridization.

The continuing researches by the Indiana group are important in another respect. When Mark G. Nielsen and his colleagues examined the genetics and molecular biology of what was happening in these hybrid experiments, comparing the hybrid offspring with the parental species, they found higher levels of paternal gene expression in the development of the hybrid embryos than have been reported from other hybrid sea urchins.[9] To put it simply, the development of the hybrid embryos was as dependent on the paternal genetic inheritance as the maternal. Moreover the genome appeared to choose paternal or maternal genetic inheritance in a nonrandom fashion, as if it could recognize the parent of origin and would choose paternal for certain developmental pathways and maternal for other developmental pathways. The researchers also found a great deal of genetic creativity in the hybrids, including new sites of genetic expression, genes with new additive expression, and novel regulatory gene interactions. This had important developmental implications at cellular and organizational levels. In sea urchins, as in insects, the cells of the skin layer, or ectoderm, play a crucial role in body form and development. The ectodermal cell types are different in the two parental species, but these merged to an intermediate form in the hybrids. Extraordinary as it might seem, the merger of the two very different parental genomes produced, in one fell swoop, a new evolutionary entity—with a different genome and different developmental pathways from either of its parents.

Nielsen, Raff, and their respective colleagues are cautious in extrapolating their findings to a major evolutionary role for hybridization because, in nature, the hybrid offspring, with its potential for creating new species, might or might not survive. Two factors might help determine its survival or extinction. It might have to compete successfully with the parental species for survival in the same ecosystem, or it might have to possess some new properties that enabled it to colonize ecological niches not habitable by either parent. We also need to consider the likelihood, as suggested by Williamson as part of his larval transfer theory, of repeated hybridization events between the same two parental species over the vastness of geological time. Repeated crosses, in different seasons, or in varying ecologies, would make it rather more likely that, from all such trial and experiment, hybrid crosses would ultimately find successful niches.

In 2002, Professor Meiko Komatsu and his graduate student, Takako Chimura, at Toyama University, Japan, reported interspecies and interclass hybrid studies involving the sand dollar *Peronella japonica*, which metamorphoses through a non-feeding two-armed pluteus, and four different species of starfish from two separate genera.[10] The various permutations of male and female cross-fertilizations resulted in six successful hybrid pairings, some involving cross-species fertilizations and some apparently cross-class. In every crop of successful hybrid offspring, some successfully metamorphosed to juveniles. This was the first demonstration that cross-class fertilizations involving these marine organisms could result in normal juveniles. Komatsu and Chimura were sufficiently encouraged to undertake further studies.[11]

As its name suggests, the sea urchin *Hemicentrotus pulcherrimus* is a beautiful creature that shares the coastal waters of Japan with a large yellow-and-purple-colored starfish known as the Northern Pacific sea star, *Astarias amurensis*. Komatsu and Chimura went on to conduct hybrid crosses between these two classes of echinoderms, with the sea urchin providing the eggs.[12] These crosses resulted in offspring that developed to pluteus larvae, which in turn metamorphosed to juveniles. But in this case genetic analysis suggested that the larvae and juveniles were not the result of true crosses but had developed from the maternal egg alone, a single-parent or "parthenogenetic" development. The pluteus larva was what would be expected from the maternal echinoderm—findings that signaled the need for caution in any interpretation of hybridization that doesn't include detailed genetic confirmation.

Nevertheless, after decades of skepticism, these observations, together with growing evidence for the importance of hybridization in many scientific departments worldwide,[13] are confirming the importance of hybridization as an important mechanism for hereditary change in evolution.

— 19 —
A New Life-Form

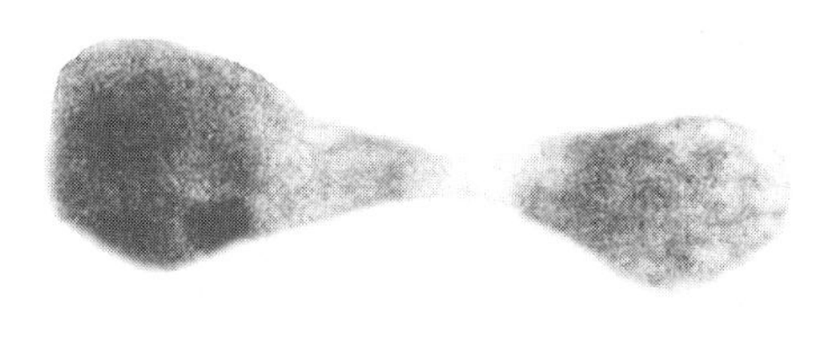

Budding hybrid spheroid

In 1996, the aspiring marine biologist Sebastian Holmes happened to share the same office in the Port Erin Marine Laboratory with Don Williamson. Holmes had recently graduated in biology at Bangor University, North Wales, where he had developed an interest in community interactions among the organisms that inhabit the rocky shore. This led him to a more specific interest in the interactions between mollusks, such as limpets and periwinkles, and the other life-forms in the shore community. Now, for his PhD, he had come to the Isle of Man to study barnacle populations, and this brought him into daily contact with Williamson, who explained the nature of his larval transfer theory. As a result of his stroke, Williamson had been unable to continue practical experimentation, but he had not abandoned his theory. "At the time, I thought, 'that's really smart. I like that,'" Holmes later recalled of Williamson's theory. "And I just logged it away in my mind."[1]

Where Williamson's interest was in larval evolution, Holmes was fascinated by the potential of hybridization to form completely new animals. It seemed to him that if hybridization was a genuine force in nature, it would give rise to sudden large-scale genetic novelty, speeding up the evolutionary process. The hybridization experiments of Raff and his colleagues seemed to confirm exactly that—and they were only crossing related species of sea urchins within the same genus. Hybridizing animals from different orders, classes, and possibly even phyla, if it proved successful, would take this idea a major step further. In the years that followed, Holmes kept in touch with Williamson, hoping that at some stage they might have the opportunity to work together.

Holmes completed his PhD in 1998 and then took a postdoctoral fellowship at the Netherlands Institute for Sea Research, where he extended his research to marine population genetics. A few years later, while waiting for consideration of further funding from the Dutch National Environmental Research Council, he thought again about Williamson and his extraordinary theory. It occurred to Holmes that he was now in a position to help Williamson repeat his hybridization experiments. All they needed was some funding. I happened to mention this in passing to Lynn Margulis, who had first alerted me to Williamson's work and theory. She generously offered Holmes and Williamson some of the prize money she had just been awarded for scientific achievement.[2] After the long and frustrating gap enforced by Williamson's disability, the two marine scientists were in a position to make plans for a new experiment.

At the time, I asked Holmes what proof he would need from the forthcoming experiment to convince the world of science that Williamson's theory should be taken seriously. "My take on Don's work," he replied, "is that he has actually done it already. He has shown that the cross-phyletic hybridization works. I would repeat his experiment, backing it up with some molecular biology. I think that once you get two different phyla to cross and show that the offspring survive, that's it. For me that would be sufficient proof. I would also think that for the general scientific community that would be sufficient. Others could investigate it further. People could go about fine-tuning it."[3]

In September 2002, Holmes, accompanied by a biology student

named Nic Boerboom, arrived at the marine laboratory in Port Erin, aiming to repeat the experiments Williamson had conducted from 1989 to 1990. They brought some useful equipment with them from Holland, including a photomicrography computer setup that would enable them to capture microscopic images of any putative hybrid offspring and their development. Throughout September and October, they collected mature specimens of the green sea urchin, *Psammechinus miliaris*, by hand from Gansey Point, Port St. Mary; they collected specimens of another sea urchin, the so-called sea potato, or *Echinocardium cordatum*, from Derby Haven; they harvested from sand at a depth of a few meters in Port Erin Bay the common starfish, *Asteria rubens*, a familiar orange species found in almost every marine ecosystem around the British Isles; and they collected the sea squirts, *Ascidiella aspersa* and *Ciona intestinalis*, from the seawater storage tank back at the laboratory. To complete their preparations, they trawled from a boat at various locations off the island to collect the "bloody Henry" starfish, *Henricia occulata*, and the spiny starfish, *Marthasterias glacialis*. All these animals were brought back to the lab, where they were washed with filtered seawater and maintained in aerated seawater that had also been filtered. Only now did Holmes, Boerboom, Williamson, and his wife combine forces to begin harvesting the eggs and sperm from this array of animals.

Seven reciprocal crosses were attempted, making fourteen hybridization experiments in total. Control experiments were set up against all of the potential hybrids. Ten involved crosses between different phyla and four between different orders. The labeled eggs were placed in a beaker of seawater and just enough of the appropriate sperm was added to tinge the water milky. After twenty minutes, the excess sperm was washed from the now fertilized eggs and these were split into separate containers and allowed to develop. Any eggs that showed no signs of developing were discarded.

Ten of the fourteen crosses, consisting of five reciprocal groups, resulted in successful fertilizations. In two of the five reciprocal groups, one at cross-phyletic level and one at cross-order level, cell division halted before the stage of a blastula, and the cells disintegrated. Of the remaining three successful fertilizations, all at cross-phyletic level, two proceeded to the development of larvae, with the larvae adopting the forms expected of the echinoderm (starfish or sea urchin) parent as opposed to sea squirt parent regardless of whether the echinoderm was maternal or paternal. But in the majority of these larval offspring, development also stopped, usually between 72 and 96 hours, causing the larvae to disintegrate.

The researchers obtained long-term offspring survivors in a single cross-phyletic hybrid, involving eggs from the sea urchin *Psammechinus miliaris* and sperm from the sea squirt *Ascidiella aspersa*. In this experiment, development proceeded to the formation of an early four-armed pluteus larva—in other words, it followed the maternal echinoderm pattern. This was a similar cross-phyletic grouping to that of Williamson's 1989 and 1990 experiments, but it reversed the parentage of the two contributing species.

In this more successful cross, the sea squirt sperm had been pipetted from a visible cloud in seawater emanating from an isolated individual—this meant that the sperm was less concentrated than in the earlier experiments, closer to the concentration that would normally be found during sexual reproduction in the oceanic environment. The three scientists followed the methodology established by Raff and his colleagues, washing the sea urchin eggs in acid seawater for forty-five seconds, rinsing them, and then fertilizing them with ascidian sperm. Controls were also set up, involving normal fertilizations of *Psammechinus miliaris* eggs with *Psammechinus miliaris* sperm. These developed into eight-armed plutei, the normal larva of the parental species.

The great majority of these hybrid eggs divided, hatched as ciliated blastulas, and developed to four-armed plutei two weeks after hatching. During the third week, however, the hybrid plutei retracted their four arms and became spheroids. These soon settled to the bottom of the culture plates, but they lacked the organs of attachment seen in the spheroids during the 1990 experiments. The new spheroids were about 0.1–0.2 millimeters in diameter, and, lacking cilia, they could move at no more than 1 millimeter per hour over the bottom of the dish. Now and then, in dramatic fashion, these reproduced, the body elongating, developing a constricted ring in the middle, and they divided, through a visible and frequently photographed budding, into two free-living individuals. Many spheroids produced clumps of cells of up to 2 millimeters in diameter, showing what appeared to be possible differentiation into different tissues, and three of these went on to resemble an early stage in the development of a juvenile ascidian, but they failed to develop into adults. Over hours, or days, they became indistinguishable from the hundreds of other structureless clumps of cells.[4] Meanwhile the spheroids went on living and reproducing by budding for many months.

Although the results were fascinating, Williamson was left somewhat disappointed. He had hoped that the hybrids would go on to produce viable and fecund adults closer in complexity to the parental generations. But for the neutral observer, the success of the cross-phyletic fertilizations and the reproduction, through budding, of the new spheroids is of immense interest.

Holmes had stated before conducting the experiment: "I think that once you get two different phyla to cross and show that the offspring survive, that's it. For me, that would be sufficient proof." And we might also recall that Richard Strathmann, at the University of Washington, had also suggested that Williamson only needed to achieve plausibility, not concrete proof.

In 2003, Holmes sent me more than a hundred photomicrographic images of the hybrid offspring, taken at each stage in their development from fertilized eggs, through plutei, and then spheroids. I gazed, astonished, at images that captured the spheroids budding. Larvae do not reproduce. Budding is a form of asexual reproduction that is seen in colonial marine invertebrates, such as corals and moss animals. It is not seen in sea urchins, such as *Psammechinus miliaris*, which had provided the eggs for the successful hybrid cross. And while budding is a feature of some colonial forms of sea squirts, it is unknown in *Ascidiella aspersa*, which provided the sperm for this hybridization.

The complex genomic reorganizations that follow hybridization are only poorly understood at present.[5] This latest cross-phyletic experiment, in merging two such radically different life-forms, was, to put it mildly, adventurous. The sea squirt, with its tadpole larva, has a tenuous link to the vertebrates, while the sea urchin belongs to a phylum that is widely separated from that of the sea squirts both on the tree of life and by separate evolutionary trajectories over hundreds of millions of years. If Williamson is right in his larval transfer theory, the larval forms of sea squirts, and thus the metamorphic life cycle, arose hundreds of millions of years ago from a hybrid fusion of two dissimilar species. But this is presumed by Williamson to have taken place at a time when animals were at a purportedly more malleable stage of their evolution, before natural selection had honed their genomes to the purportedly less malleable modern stage.

Other biologists have produced many successful laboratory hybrid experiments among butterflies, frogs, fish, and marine invertebrates, and hybrids are being increasingly observed in nature, where success of the offspring usually results from crosses between closely related species. But the fusion of genomes from different phyla must surely result in an initial genomic chaos in the hybrid offspring, deriving from the coming together of hundreds, perhaps even thousands of pre-evolved genes from two very different evolutionary pathways.

It seems likely, from what little we do know of the genomic responses to hybrid crossing, that complex internal mechanisms will have attempted to rationalize the potential outcomes. Key to any understanding of this very complex situation are the developmental pathways that control not only form but also the differentiation and development of internal structures, limbs, organs, tissues, and specific cells. Such developmental pathways will themselves be controlled not only by genetic factors but also by epigenetic controlling mechanisms—mechanisms involving complex chemical control of the expression of genes—that are capable of responding rapidly to change in the genetic environment.

Various outcomes are possible in such cross-phyletic crosses. Commonly, and most frequently seen in the experiment above, there will be complete failure of development. Another possible outcome, amply demonstrated in Raff's experiments in much more closely related hybrid crosses, is complex change in genetic regulatory pathways, with suppression of some pathways, enablement of others, and, most tantalizing of all, the creation of new developmental pathways. All of this would suggest that unknown genomic determinants and constraints might dominate at different times and different situations, sometimes selecting pathways of maternal origin, sometimes pathways of paternal origin, and at other times blending the two disparate inheritances.

Could the results of the successful cross-phyletic experiments be a fluke, based on contamination with pre-fertilized echinoderm eggs? None of the three scientists who participated in the experiments view contamination as a likely possibility. Even if we discount their expert witness, accidental contamination would have most likely resulted in just an occasional success. Here the great majority of the eggs were observed to develop in a similar if wholly novel pattern. This makes the possibility of accidental contamination very unlikely. Another rather unusual possibility, mentioned earlier in the Japanese experiments, is a spurious development from the egg alone, a form of parthenogeneis known in the scientific jargon as "gynogenesis." This would imply that the sperm, while triggering development, failed to merge its chromosomes with the maternal chromosomes in the egg. The subsequent offspring would be programmed entirely by the maternal genome leading to a single-parent offspring, or parthenogenesis. But an explanation based on gynogenesis would need to explain why, if development took place from the maternal genome, none of the plutei developed beyond the four-armed stage, why all resorbed their arms to become spheroids, and why some of these spheroids appeared to attempt a further development to the paternal ascidian body form. It would also need to explain how many of the hybrid offspring entered a final phase of budding reproduction. There is no known biological, or genetic, explanation of how a development, based wholly on the maternal sea urchin genome, would give rise to a budding spheroid.

It would appear plausible that the offspring of the Holmes, Boerboom, and Williamson experiments were true cross-phyletic hybrids. If so, there is also an additional astonishing potential implication of this research.

The spheroids, in their capacity to reproduce through budding, could be seen as a new, albeit artificially created, life-form—possibly the first such lab-created life-form. This in turn poses a number of interesting questions. Is it possible, for example, that the only viable offspring from such a disparate genomic union was an exceedingly simple one, a solution that, perhaps through massive suppression of developmental pathways, rediscovered the most basic of all animal forms—one that physically, and reproductively, resembled the nearest thing to a common ancestor of the two distantly related marine invertebrate phyla?

In the words of Scott F. Gilbert, "one of evolution's most important experiments was the creation [from single-celled ancestors] of multicellular organisms."[6] Ernst Haeckel proposed that the ancestor of the entire animal kingdom was a gastrula-like ball of cells with a simple gut, which he called the "gastraea." But there is an obvious alternative: an even simpler organism, such as the simplest of all larval forms, the spheroid-like "planula," which is the larva of jellyfish and medusas. The planula has no gut but imbibes nourishment through its surface layers. For Balfour, the planula was not merely the earliest example of a primary larva, it also represented the primal form of the common ancestor of the animal kingdom.

Forms resembling fossilized blastulae had been discovered in Chinese and Siberian rocks that may represent embryos from the very earliest animals, dating to between 500 and 600 million years ago.[7] Yet these may be eclipsed in age by much earlier spherical forms in fossils from the Stirling Range in southwestern Australia that may have been the oldest ancestors of multicellular life, such as animals and plants, dating to 1.2 billion years ago.[8] Other paleobiologists have reported equally ancient fossil forms that resemble embryos and planula larvae,[9] though this has been disputed by others.[10]

Williamson's extraordinary cross-phyletic hybridization experiments deserve to be repeated by colleagues equipped with state-of-the-art molecular and genetic technology, so that the outcome can be assessed with the same intensity of scrutiny, at every stage of development, as seen in the work of Raff and his colleagues. How stunning it would be if further experiments, backed up also by epigenetic study of developmental sequences, were to confirm that Williamson's spheroids are not merely hybrids across different phyla, but might also be a physical recapitulation of our earliest multicellular ancestors from so very long ago!

PART FOUR

The Molecular Age

After the structure of DNA was solved . . . again and again, the smallest, most casual beginnings—a skeptical question asked in a Paris café, some specks seen on an electron micrograph, the idea struck off in a sentence during a drive down from New England—have grown up into specialties that have engrossed entire lifetimes in science, today command whole teams and laboratories, whose results fill volumes and are not ended.

—HORACE FREELAND JUDSON,
The Eighth Day of Creation

20

The Puzzle of the Hornworm Brain

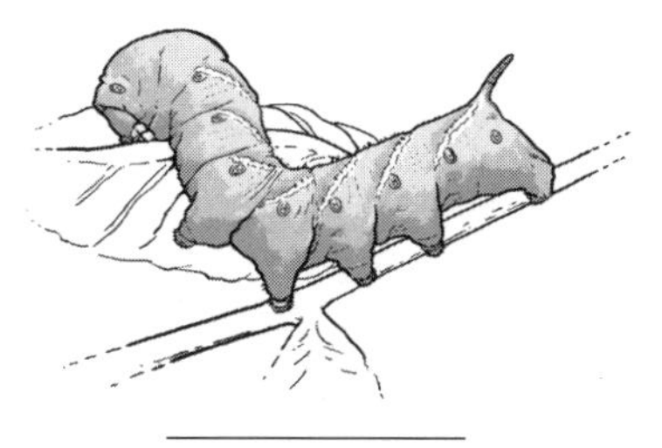

Tobacco hornworm

On September 30, 1999, the publication of a scientific paper in *Nature* caused a flurry of excitement that spilled over into the national media in America. The title of the paper was "The origins of insect metamorphosis,"[1] and the work of its authors, James W. Truman and Lynn M. Riddiford, based in the Department of Biology at the University of Washington, extended the earlier researches of Vincent B. Wigglesworth, Carroll Williams, and the many other experts involved in the search to understand metamorphosis. The reason for the excitement was that the paper questioned some of the conclusions drawn by the earlier scientists.

We saw earlier how Wigglesworth's interest in metamorphosis lay not in its evolution but in its physiological control. While juvenile hormone and ecdysone provided a vital understanding of the physiological control mechanisms, and must surely have played an important role in the evolution of metamorphic change, attempts to extrapolate this to other mysteries, notably pupation, remained unsatisfactory. Like Darwin, Wigglesworth had played down the differences between complete and incomplete metamorphosis, assuming that the dramatic, even cataclysmic, changes seen in complete metamorphosis—changes that appeared to mirror those of many marine invertebrates—should be seen as exceptional. Both scientists assumed that the more gradual change as seen in the incompletely metamorphosing insects was more representative.

Lynn Riddiford had coauthored a number of earlier papers with Carroll Williams,[2] and both she and Truman had worked with Williams at his laboratory at Harvard. But now they were exploring metamorphosis from a different angle than that taken by the earlier researchers. Where Wigglesworth, Willams, and their colleagues were largely interested in the physiology of metamorphosis, Truman and Riddiford were interested in a new avenue of biological thinking, part of a field that brought together the disciplines of evolutionary biology and development into a single discipline—evolutionary development, or "evo-devo." The youngest of the evolutionary disciplines, this was granted its own division in the Society for Integrative and Comparative Biology (SICB) in that same year, 1999.[3]

Evo-devo studies how embryological or post-embryological development affects evolution. This had obvious relevance to the evolution of metamorphosis, which is essentially post-embryonic development. In a recent review of how evo-devo works,[4] Sean B. Carroll makes the very relevant point that two-thirds of the gene sequences exposed to natural selection in the human genome are non-coding—they don't translate into proteins that form part of the physical, or chemical, structure of the organism. The existence of this vast, mostly unknown genetic territory might lead to unrealistic expectations about what can be learned from comparisons of protein-coding genome sequences alone. In fact a great many genes play a covert role in coding for "regulatory elements"—genetic sequences that direct embryological development, or, in adult life, the expression of protein-coding genes. The origins and functions of these "regulatory" sequences are of major evolutionary importance.

In their 1999 paper, Truman and Riddiford reappraised insect metamorphosis in a series of logical steps. "Metamorphosis," they explained, "is one of the most widely used life-history strategies of animals." They agreed with the principle, espoused by the eminent Danish zoologist Claus Nielsen, that differences between the larval and adult forms allowed the same organism to exploit very different habitats and food sources.[5] In marine invertebrates we have seen how metamorphosis allows a single organism to take advantage of planktic dispersal as a larva and then food resources and establishment of territories as a bottom-dwelling adult. In amphibians, it allows the organism to take advantage of marine and terrestrial habitats, and in completely metamorphosing insects, it allows the adaptation of the plant-eating grub or caterpillar larva to the winged adults, which are often non-feeding, the latter providing sexual reproduction as well as assisting dispersal to the widest possible geographical range. Truman and Riddiford also disagreed with Williamson in sharing the view of the Scandinavian experts, Gösta Jägersten and Nielsen, that metamorphosis in amphibians and many marine invertebrates was an integral part of the animals' earliest life cycle.[6]

But this did not apply to insects. We have already seen that the first insects, which originated in the Silurian some 438 to 408 million years ago, were wingless and did not undergo metamorphosis. This means that metamorphosis must have been introduced into the life cycle through some later evolutionary adaptation. As Truman and Riddiford now explained, insects with incomplete metamorphosis—for example, cockroaches and dragonflies—had mixed evolutionary origins. In other words, incomplete metamorphosis evolved again and again, such multiple origins, in the scientific jargon, being "polyphyletic." We might also recall, from the example of *Rhodnius prolixus*, that the larval equivalents among incompletely metamorphosing insects look like immature adults and are known as nymphs. Besides lacking genitalia, nymphs have wing buds folded over their backs, and these wing buds transform into the fully functional wings during the final molt to the adult. Truman and Riddiford now highlighted some important differences in the evolution of incompletely and fully metamorphosing insects.

"Insects with 'complete metamorphosis' were first seen in the

Permian 286 to 245 million years ago and constitute a monophyletic group"—a group with a single evolutionary origin. This suggested a single evolutionary event gave rise to the wonder of butterflies, beetles, moths, flies, and bees.[7] In emphasizing this difference in the origins of incompletely and completely metamorphosing insects, Truman and Riddiford brought into question the central assumption Wigglesworth had made at the very beginning of his researches on metamorphosis—an assumption that by and large had been accepted by the world of entomology—that his findings about the incompletely metamorphosing bug *Rhodnius* would be equally relevant to fully metamorphosing insects.

This physiological perspective had proven useful to Wigglesworth, enabling him to simplify his experimental designs. But from the evolutionary perspective, and in particular that of complete metamorphosis, with the catastrophic changes within the pupa, the physiological approach left key aspects that still needed to be explained. How, Truman and Riddiford wondered, did complete metamorphosis evolve from the simpler life cycles of the ancestral insects? This demanded further investigation. Perhaps most astonishing of all, in now questioning the prevailing viewpoint they reopened a controversy about metamorphosis that began some 2,300 years earlier with the beliefs of Aristotle, who had, as we might recall, put forward the theory that the caterpillar was nothing more than a soft egg.

It all began with a romantic story. Lynn Riddiford first met her future husband, James Truman, at Harvard, in 1967, soon after the younger Truman began as a graduate student in the department of Carroll Williams. In Truman's words, "As it turned out, I was assigned Lynn as my graduate adviser. So I got my degree working with her. After I got my degree, we were married, and I stayed on for an additional three years in the department." This meeting of two like-minded personalities would give rise to one of those wonderfully creative partnerships in science that have been described as the marvelous resonance between two minds.[8]

Riddiford did not begin her scientific career with an interest in metamorphosis. Originally her interests ranged over a wide variety of subjects from physical chemistry to the ability of newts to regenerate lost limbs. But she switched to metamorphosis when she was an undergraduate working with Williams. "He'd just given us a talk about juvenile hormone, which I had never heard of before. I asked him the odd question or two. For example, I wanted to know if juvenile hormone had anything to do with the metamorphosis of tadpoles. He said he didn't know—but would I like to try to find out? I investigated it and found that it didn't have any effect. But anyway, that's where my interest began."[9] Truman, on the other hand, had always been interested. He had become an avid insect collector while still in his early teens, inspired by Fabre, and devoured the French naturalist's books. As an undergraduate at the University of Notre Dame he became involved with mosquito research before he moved on to his graduate education in Williams's department at Harvard.

I asked him, "What was it about insects that so interested you?"

"It was the diversity. I was a bit of a morphology freak. In one sense, insects have a very simple body plan—but, thanks to this amazing cuticle, they can make such a diversity of shapes and patterns. I went through a butterfly phase. But over time butterflies just didn't excite me that much any more. Morphologically, they're quite similar, although the colors are beautiful, quite fantastic. I was far more interested in beetles, because of the huge morphological diversity." Darwin had arrived at the same conclusion. Amazingly, there are about half a million species of beetles with myriad varieties of shapes, colors, and living strategies.

While working for his PhD in Riddiford's lab, the youthful Truman made his first discovery. The hormonal control mechanisms of metamorphosis were already established by then, but there was no hint that the behavior of insects at the time of molting might also be controlled by a hormone. Truman's earliest work, under the tutelage of Riddiford, was with the giant silk moths favoured by Williams, but in time he switched his experimental subject to the sphinx moth, *Manduca sexta*, whose larve is the tobacco hornworm. One of the familiar garden pests in America, the hornworm is pea green in color with white stripes running along its sides, and, growing to a gigantic four inches in length, it can quickly defoliate crops such as tobacco, tomatoes, potatoes, eggplants, and peppers. It is also recognizable from the prominent red spike at its rear, curved like a horn, that gives it its name. When, under Riddiford's supervision, Truman adopted this giant caterpillar as the subject for a series of investigations of molting behavior, he discovered that a brain-sourced hormone—"the eclosion hormone"—was responsible for larval behavior at the time of molting. Over subsequent decades, the two biologists went on to locate the cells in the insect brain that secrete the hormone, its amino acid sequences, and finally the gene that codes for its sixty-two amino-acid peptide sequence.[10]

Today we know that the eclosion hormone is released during a precise stage of the molting cycle, when it acts directly on the brain to trigger the molting behavior, leading in turn to tissue changes and the behavior that enables the shedding of the old cuticle.

The eclosion hormone appears to be unique to insects, and the diversity of its effects are still being evaluated. Finding the eclosion hormone was just one of many important discoveries made by these two remarkable scientists. The range of their investigations covers many different aspects of insect development, metamorphosis, and evolution. In one investigation, for example, the two scientists looked at the hormonal control of migratory flight in the large milkweed bug.[11] Studies such as this are now helping us understand the annual monarch butterfly migration, which, in Truman's opinion, is most likely cued by the shortening period of daylight, and probably linked to changes in juvenile hormone secretion.[12]

In the study of development, most of their fellow entomologists had been content to accept that the mystery of insect metamorphosis had been solved by the discovery of the controlling hormones. But Truman and Riddiford, who had worked in the laboratory of one of the pioneers, felt differently. In their subsequent scientific careers, they accumulated a vast knowledge and experience of insect development, and, by implication, insect evolution. In the course of their research, however, they encountered numerous instances of developmental phenomena that appeared to make no sense according to the prevailing theories of insect metamorphosis. In Truman's words, "This became glaringly evident from the [development of the] nervous system."

In the mid 1980s, Ron Booker, a postdoctoral researcher working under Truman's guidance, was studying the development of the nervous system in the caterpillar of the sphinx moth. In nerve clusters known as the thoracic ganglia, Booker came across stem cells that generated large numbers of new neurons as the larva grew, but these neurons remained dormant in the caterpillar, only coming to maturity during the metamorphosis to the adult moth, when they produced most of the adult nervous system. Stem cells that develop into nerve cells are known as neuroblasts. A little later, when Truman and Riddiford took a sabbatical to work in the lab of Mike Bate at the University of Cambridge, Truman found similar stem cells in the larva of the fruit fly. This confirmed that such neuroblasts were of critical importance to the metamorphosis of the moth and the fly. Truman wanted to know more about them. Where did they come from? "Were they a new set of stem cells unique to larvae that undergo complete metamorphosis? Or were they stem cells that had persisted from the embryo stage—in other words, were they embryonic stem cells that had lain dormant in the embryo only to activate in the larva to make the rest of their lineage of neurons?"

Truman knew that in incompletely metamorphosing insects, such as grasshoppers, similar stem cells generate all of the neurons in the embryo prior to hatching, after which they appear to die off, so that, by the time these incompletely metamorphosing varieties reached the nymphal stage, their ganglia contained no residual neuroblast stem cells. It suggested a major difference in the metamorphic development between completely and incompletely metamorphosing insects.

After the sabbatical at Cambridge, the partners traveled to Nairobi, where they turned their attention to tsetse flies, the insects that carry the trypanosome parasite of African sleeping sickness. Tsetse flies, like sphinx moths and fruit flies, undergo complete metamorphosis. Again the researchers found those same stem cell neuroblasts, which they could track from their appearance in the early embryo and into the larva, when they switched to making the neurons for the adult version of the nervous system. These discoveries were confirmed by some elegant transplant experiments in the fruit fly performed by researchers elsewhere.[13] By now Truman and Riddiford were convinced that they were seeing a pattern that pointed to an important difference between complete and incomplete metamorphosis in insects.

This conviction was at the back of their minds when, in 1996, they traveled to Australia to spend several months working with Eldon Ball at the Australian National University in Canberra.

— 21 —

Aristotle or Darwin?

Pronymph of milkweed bug

In fact it was their second trip to Australia—three years earlier Truman and Riddiford had worked, respectively, at the Australian National University and the Commonwealth Scientific and Industrial Research Organisation (CSIRO), when Truman's interest was insect embryos and Riddiford's continued to be the fruit fly. This earlier experience had given them a better appreciation of the embryonic development of simpler insects, such as grasshoppers.

This second Australian trip came as a breather from their normal routines, allowing them time and opportunity to think about where all of their work was heading. In spite of the anomalies, they still accepted the prevailing views of Wigglesworth and Williams. But they also knew that, somehow, the wonder of metamorphosis had interposed itself into the life cycle of insects that had already evolved to colonize dry land without the need for, or advantages of, metamorphosis. At the same time the two researchers were increasingly questioning the generalization that lumped completely and incompletely metamorphosing insects into the same evolutionary, and developmental, explanation. Even today four of the most primitive orders of insects are wingless and still do not metamorphose, including the bristletails and the springtails, which include the earliest fossil insects. Now, in Truman's words, they arrived at the decision: "Maybe we should look at the embryonic development of the more basal insects."

They turned their attentions to two familiar examples, silverfish and locusts. Silverfish belong to the order Thysanura—the bristletails mentioned above—a primitive group of wingless insects that live in soil, rotting wood, or the crevices of kitchens and bakeries. They still retain rudimentary appendages on their abdominal sections that may be a survival from still more primitive forms that had limbs on all of the body segments, rather as we see in millipedes and centipedes. They grow through molting, but do not undergo metamorphosis. Locusts are members of the order Orthoptera, which includes grasshoppers and crickets, with fossil members that date back to the Carboniferous period, some 360 to 286 million years ago, and which undergo incomplete metamorphosis.

Truman and Riddiford wrapped batches of eggs of these two insect species in small pieces of cheesecloth and dropped juvenile hormone, dissolved in a little acetone, on top of them. They didn't need to perforate the egg membrane because the hormone readily diffused through the membrane and entered the embryo. Some of these eggs were transparent, so they could transilluminate them and photograph what was happening inside the membrane. The results were startling: "Suddenly," in Truman's words, "we began to see that a lot of the issues we had been concerned about in metamorphosis came into play in the embryos of these more basal insects. That made us start to think afresh about the current notions people were holding."

The non-metamorphosing silverfish and incompletely metamorphosing locusts hatch out from the eggs as nymphs, which closely resemble the fully grown adults. Wigglesworth, who more than any other had established the paradigm of insect development, had assumed that the larvae of fully metamorphosing insects were equivalent to the nymphs of non-metamorphosing and incompletely metamorphosing insects. The prevailing modern hypothesis further assumed that, as the disparity between the larval and adult stages of fully metamorphosing insects grew—through the action of natural selection on the two very different stages and life cycles—there was a biological need to "bridge" the widening gap in order for the adult to emerge. This bridge, as a result of evolution over time, became the pupa. A purist might balk at the concept of need, since evolution is a chance affair paying no heed to human concepts of need. An evolutionary biologist might better express this as "selection pressure for a transitional form." But this was a hypothesis, however expressed, that Riddiford and Truman now felt compelled to question. An alternative explanation that might better fit the evidence lay hidden, largely forgotten, in the murky pages of history.

Aristotle believed that the caterpillar was a soft egg—in other words a continuation of embryonic development. In 1651, at Wigglesworth's cherished Caius College, Cambridge, William Harvey took Aristotle's thinking further, supporting his idea of the caterpillar as a free-living embryo, but suggesting that the pupa, and not the caterpillar, of the butterfly was the equivalent of the nymph of an incompletely metamorphosing insect, such as a dragonfly. In 1913, the Italian naturalist Antonio Berlese extended this thinking, proposing that metamorphosis arose through the hatching of the insect larva at a premature stage of embryonic development—a view known as the de-embryonization theory.[1] In the first half of the twentieth century Berlese's theory was further developed by J. J. Jeschikov, who pointed out that there was a close link between the amount of yolk in the insect egg and the timing of hatching.[2] If there was less yolk, say in the butterfly egg, the larva hatched out early as a caterpillar. For a time, the de-embryonization theory, also known as the Berlese-Jeschikov theory, captured the imagination of biologists.

Wigglesworth disagreed with these and other alternative theories.[3] Darwin had earlier proposed that metamorphosis was the development of a single life-form embracing a number of different body forms, or polymorphisms, so that each body form was a step, or plateau, along the developmental pathway. This allowed natural selection to adapt each stage along lines that fitted the selective pressures of that stage. Wigglesworth agreed with this Darwinian explanation: "The most likely supposition is that there has been an independent evolution of the different 'stages' of insects."[4] This organism had somehow evolved the ability to change its form at different periods of its development. He believed that insects contained in their genetic makeup the potential for these different body forms. This does not imply that, in the language of science fiction, they were genetically controlled to be shape-shifters. Rather it presaged the modern evo-devo interpretation: the different body forms were controlled by the genetic makeup of the complex organism operating through a series of discretely inherited developmental pathways. Given such a perspective, we can see why Wigglesworth saw no essential difference between the process, and its evolution, in all metamorphosing insects. The incompletely metamorphosing insects had evolved the potential for nymph and adult forms, whereas the fully metamorphosing insects had, through very similar mechanisms, evolved the potential for larva, pupa, and adult.

If Wigglesworth and Darwin are right, this explanation is as good an explanation of the evolution of metamorphosis as we're going to get. The evolution to complete metamorphosis, a more gradual process in incomplete metamorphosis, evolved the means of wiping the slate clean so that, within the pupa, it could start all over again from "set-aside" stem cells to effect the phoenixlike rebirth. To Truman this hypothesis seemed too conceptually nebulous—and, moreover, it failed to explain the anomalies that now seemed glaring in light of his thinking and experience. In attempting to formulate an answer to the problem of the evolution of insect metamorphosis, he considered the alternative theories, including those of Harvey, Berlese, and other successors to Aristotle's view. In his words, "The essence of the Berlese theory—as with Harvey and Aristotle—is that the larva is a feeding embryo, or [to paraphrase Aristotle], a feeding egg. That's very nice, but the real issue is what really happens if an embryo hatched prematurely in this way? The same thing would apply to a baby born prematurely. The maturation isn't finished and the animal would not be viable. Therefore, if you are going to [propose a theory in which] an animal hatches at an earlier embryonic stage, [you're going to have to explain how it has come to] be developmentally functional."

Truman and Riddiford thought it worthwhile probing the evo-devo conundrum.

In metamorphosis, juvenile hormone is a "status quo" hormone, a term coined by Truman's former professor, Carroll Williams. It blocks development, keeping the insect in the same form as it was prior to a molt. But now, Truman and Riddiford found that, in the silverfish and locusts, there appeared to be an altogether different role for the same hormone while the embryo was still forming within the egg. "When we added juvenile hormone during embryonic development, we found that instead of keeping the embryo in status quo—as an embryo—it actually caused the embryo to undergo premature maturation." This was a startling revelation.

When the two investigators went back and hit the journals in the university library, they found that it should not have come as such a great surprise. There was already some information in the scientific literature to suggest that the embryonic effects of JH were different in basal insects when compared to more advanced insects. People had simply ignored these earlier findings, leaving it for Truman and Riddiford, in a sense, to rediscover the wheel. But as Truman expressed it to me, "It also involved our looking at the wheel in a different way." They realized that this discovery had potential evolutionary significance and so probed it further.

When they treated silverfish eggs with juvenile hormone, it had a devastating effect on the development of the embryo. In the insect embryo, the limbs begin to grow out from appropriate segments along the body axis. Normally they begin as extensions from the body, and then they segment, to give rise to the typical multi-jointed insectile legs. This is followed by the laying down of the mature covering layer, or cuticle, and by the development of the appropriate muscles of attachment. When Truman and Riddiford added juvenile hormone to developing embryos of silverfish—say, right after they first developed limb buds—the embryos laid down the mature cuticle on the buds, halting further development of the legs. The exposure to juvenile hormone caused a premature maturation, which, in the case of the silverfish, resulted in an embryo that was not viable. This implied that during the normal embryology of these basal insects, juvenile hormone would not normally come into play until the very end of development. When, on the other hand, they repeated the experiment with more advanced insects, such as incompletely metamorphosing grasshoppers and crickets, they discovered that juvenile hormone had no such effect at this early stage in development. The effects of JH only kicked in when they reached a stage close to hatching. Then JH caused a premature hatching, but in these more advanced orders it took place at a mature enough stage to give rise to viable animals.

In a mammalian pregnancy, birth is very precisely timed, as is the hatching from the egg of birds, reptiles, and insects. The developmental timing of hatching is critical, as the newborn or hatchling emerges from the protected environment of the womb or egg into the less protected environment of the outside world. Nevertheless, across the spectrum of different groups of animals, birth or hatching takes place at very different stages of development, including what is clearly an intermediate developmental stage in some.[5] For example, the kangaroo young emerge while the equivalent of embryos and complete their development in the maternal pouch. In many spider groups the hatchling larva is typically non-feeding, lacks pigmentation, has limited locomotion, and continues to feed off its yolk stores; in scorpions the non-feeding hatchlings are carried for up to two weeks on the backs of the mothers. Nowhere is the hatching of offspring at embryonic stages more obvious and startling than the larval forms of the many aquatic animal phyla, with sea urchins, for example, emerging as living blastulae. Perhaps it is not surprising that a specialized hatchling stage might also be seen in the more basal orders of insects.

There is no general name given to this premature hatchling stage in insects, so Truman and Riddiford decided they would adopt the name used in the dragonfly literature—they called it the "pronymph."[6] In the life cycles of the non-metamorphosing and incompletely metamorphosing insects, they now proposed, there exists a largely ignored stage, the pronymph, which occupies a stage between the developing embryo in the egg and the nymphal stage of the incompletely metamorphosing insects. The pronymph has a number of characteristics that make it unique—its body proportions, the microscopic structure of its cuticle, its incompletely developed mandibles, and so on—in other words it is significantly different, and simpler in structure, than the nymph. They now proposed that the pronymph of these more primitive orders of insects, including the basal orders and some of the incompletely metamorphosing orders, such as the dragonflies, was the evolutionary ancestor of the larva of insects that underwent full metamorphosis. They also proposed that the nymphal stage of the incompletely metamorphosing insects had evolved into the pupal stage of fully metamorphosing insects.

The two entomologists admitted that there were residual problems with this theory. The pronymph lasts for only a single phase in the basal insects and dragonflies, while larvae go through a succession of molts and stages. But modifications of hormone control might explain these differences. Their theory, which amounted to a renaissance of the idea first put forward by Aristotle, could be seen to fit with what was already known to happen in both incomplete and complete metamorphosis. In incompletely metamorphosing insects, such as dragonflies and *Rhodnius prolixus*, the first withdrawal of JH results in metamorphosis to the adult. In completely metamorphosing insects the situation is more complex, with a small and then a larger ecdysone peak, the first priming certain tissues to get ready to change to those of the pupa, and the second giving rise to the pupal molt. JH is absent at the beginning of the pupal molt, but it subsequently reappears during early pupation only to disappear again before the ecdysone peak that gives rise to the formation of the adult insect within the pupa.

In 2004, the Truman-Riddiford hypothesis received its first molecular confirmation, when Deniz Erezyilmaz, a doctoral student working under their direction, examined the effects during development of a gene called "the broad complex."[7] Riddiford's lab had already shown that this gene controlled the expression of the pupal stage. When expressed, for example during a larval molt in the fruit fly, the larval tissues begin to activate pupal genes. Even if the broad complex was expressed in what should be the final molt to the adult, instead of activating adult genes the tissues reverted to activating pupal genes. So the broad complex appeared to specify development of the insect into a pupa. It was not activated—indeed it could not be activated—in the larval-to-larval molts of completely metamorphosing insects, such as the grub of a bee or the caterpillar of a butterfly, because it would turn the larva into a pupa. When, for the first time, Erezyilmaz examined the broad complex's role in the more basal insects, she discovered something relevant and revealing. Whereas in the fully metamorphosing insects the broad complex was only expressed in the pupal stages, in the incompletely metamorphosing insects the same complex was expressed in the nymphal stages. This offered molecular support to the evolutionary implications of the Truman-Riddiford theory, which suggested that the pronymph and nymph of incompletely metamorphosing insects were the equivalents, and thus the evolutionary ancestors, of the larva and pupa of fully metamorphosing insects.

22

Cues and Common Links

Red abalone

On January 26, 1971, while researching his book on the DNA story, Freeland Judson traveled to the University of Cambridge to interview the molecular scientist Sydney Brenner. Over dinner that evening at King's College conversation turned to the way science was changing. Brenner, one of the elect few at the center of the DNA story, commented that biologists can only be interested in three basic questions: "How do things work? How are they built? How do they evolve?"[1] Much as Wigglesworth saw physiology at the root of all questions, to Brenner's way thinking it was the second of the three questions that took precedence. It was "the deeper question that comes before you can answer the evolutionary question: How are organisms built?"

One of the wonders of nature, and closely related to the mysteries of metamorphosis, is how a single fertilized egg develops into an oak tree, a blue whale, or a human baby. Professor Eric H. Davidson, head of the Davidson Lab at the California Institute of Technology, has made a major contribution to our understanding of the genetic control of development. This has included a directorial role in unraveling the sea urchin genome.[2] In his book *Genomic Regulatory Systems*, Davidson acknowledges the importance of Darwinian evolution before going on to explain that "classical Darwinian evolution could not have provided an explanation, in a mechanistically relevant way, of how the diverse forms of animal life actually arose during evolution, because it matured before molecular biology provided explanations of the developmental process."[3] By "forms" Davidson is referring to the shapes and physical structures of living organisms. By Darwinian evolution, he is referring not to Darwin's theory itself but to the synthesis theory of the 1930s, in which, as he explains, "the argument was that organismal evolution is the product of minute changes in genes and gene products, which occur as point mutations and which accumulate little by little, providing the opportunity for selection and ultimately reproductive isolation. The major forms this argument has taken have focused on stepwise adaptive changes in protein sequence, but this is probably largely irrelevant to the evolution of any salient features of animal morphology."

In essence, Davidson is proposing that changes in the DNA sequences that control developmental processes, as opposed to changes in the DNA of genes that code for proteins, have played a fundamental role in the evolution of form. Sean B. Carroll, another evo-devo expert, makes the same point when he explains that "regulatory sequences are so often the basis for the evolution of form that, when considering the evolution of anatomy (including neural circuitry), regulatory sequence evolution should be the primary hypothesis considered."[4] The work by Truman and Riddiford on the effects of hormones on insect embryos is a perfect example of this line of thinking. Understanding the evolution of metamorphosis, including its hormonal triggers and the resulting genetic cascades, affords insight into how organisms are built—one of Brenner's key questions.

How such developmental pathways work, their biochemistry and physiology, and how they evolved, is the very essence of the story of evolution in general and the evolution of metamorphosis in particular. To learn more, we could do no better than return to the pioneering Wigglesworth and to his study of the blood-sucking insect *Rhodnius prolixus* as his experimental subject.

Adult females of this species, as we've seen, produce new batches of eggs each time they drink blood. It's the blood meal that acts as the primal signal or cue for each successive molt. There is a physiological explanation for this. Blood proteins supply the amino acids needed by the female for egg production, in particular for the manufacture of the yolk protein, or "vitellogenin."[5] As Wigglesworth demonstrated, it is the physical stretching of the larval abdomen that signals the brain to activate the hormonal triggers for metamorphosis. Juvenile hormone also stimulates yolk-protein synthesis in the insect ovary.[6] And what works for Rhodnius also works, with perhaps some variation, for other insects. For example, in many mosquitoes, egg production is also triggered by a blood meal. Only female mosquitoes feed on blood. Without the blood meal, they do not make yolk protein. In *Aedes aegypti*, the insect that spreads the yellow fever virus, the digested products of the blood meal stimulate the insect brain to secrete egg-development hormone, which in turn stimulates the ovary to make eggs.

Environmental cues are also important triggers of marine metamorphoses. These are numerous in type and can vary between species within the same genus or family. Detection of the cue, which depends on a specific sensory system, ensures that the larva settles in a habitat suitable for growth and survival, and this often implies a firm surface and a convenient food source, such as a specific alga. While some signals encourage settlement and metamorphosis, others inhibit it, so that the larva must take into account a complex cocktail of conflicting signals within a confined environment.[7]

Take, for example, the Californian red abalone known as *Haliotis rufescens*. The largest abalone in the world, and historically the most important commercial species on the West Coast of the United States, this impressive mollusk inhabits intertidal and subtidal rocky areas along the Pacific coastline, where the adults use their powerful feet to cling to rocks, allowing it to graze on kelp and other algae. The adults reach sizes upward of a foot and can live for as long as fifty-four years. The great size and the succulence of its flesh led to such overfishing that the species is now protected from commercial exploitation, although recreational fishing is still allowed off the north Californian coast. Biologists have taken a special interest in its natural history and life cycle. The mollusk's striking red color derives from the same pigment that is found in red algae. Indeed, so beautiful is the shell that Native Americans used its color, spiral form, and mother-of-pearl interior to construct "banjo-style" pendants and ornaments that were used in Kuksu dances.

Biologists have taken a special interest in the red abalone's natural history and life cycle. The mollusk reproduces through separate sexes, and it fertilizes externally. Mature females lay up to fifteen million eggs annually, which hatch into free-swimming trochophore larvae, about two millimeters in diameter. The trochophore metamorphoses to a more complex "veliger" larva, which, though only slightly bigger, develops two ciliated flaps that enable locomotion, and thus reduce the risk of being eaten. The veliger increases in size, develops more complex internal organs, secretes a protective shell, and grows what will become its future, often iridescent, foot. During this series of changes, the larval body undergoes a twist through 180 degrees on its longitudinal axis. After another two to three weeks, it metamorphoses to the juvenile adult, which abandons its planktic existence and settles to the ocean floor.

We might pause to reflect on the fact that the abalone has four distinct periods of development, first as the embryo within the egg, then as each of the two larval phases in turn, and finally as the adult. Each development requires different genetic programming, which must have evolved separately, whether through selection operating on distinct phases of development within the same organism, or, if Donald Williamson is right, through selection operating on entirely different organisms that subsequently hybridized to the single genome of what we now recognize as the abalone—or possibly a combination of both. The veliger of the Californian red abalone also settles in response to a specific environmental cue, which kicks in when it first makes physical contact with coralline red algae. The briefest of contact is all it takes. It would appear that a receptor on the larva recognizes a highly specific chemical on the surface of the alga. So vital is this environmental cue that contact with the alga triggers metamorphosis in the larva, which stops swimming and begins the final development to the juvenile adult.

Another interesting example of environmental triggering is the symbiotic incorporation of the luminous bacteria *Vibrio fischeri* into the body of the juvenile squid *Euprymna scolopes*, where the bacteria are so important that the squid cannot develop without them.[8] Symbiotic bacteria are also essential to the development of many other marine species, just as they are to insects and animals. A delightful example in the world of insects is found in the leafhopper *Euscelis incisus*, where symbiotic bacteria within the eggs are transferred to future generations. If the bacteria are killed off inside the eggs, the embryos fail to form a gut. The environmental cues that trigger and control development are not necessarily chemical. Sometimes they are physical, for example, the influence of gravity, or the effects of ambient temperature, light duration, and the humidity changes associated with the seasons.

How does the internal machinery of life respond to these environmental signals? Very likely this involves a range of very different molecular and genetic mechanisms. We know that the environmental cue initiates some key process of internal change. In plants one such response mechanism involves the non-DNA-based epigenetics systems—the non-DNA-based chemical process that control the activation of genes—that allow environmental signals to trigger gene activation in response to the arrival of spring.[9] In insects, of course, this takes the form of a hormonal response that in turn initiates the genetic cascades that program development.

Even before the discovery of the structure of DNA, scientists had observed a curious phenomenon that took place on the chromosomes of fruit fly larvae. Imagine the chromosomes as incredibly long strings of beads, each individual bead exceedingly narrow and with hundreds, sometimes even thousands, of beads being strung tightly together along a single chromosome. Today we know that these beads, which, in insects, are visible with the light microscope, represent genes. During metamorphosis, specific beads at some points along the string of the chromosomes began to swell up, so they resemble woolly scarves. This phenomenon is known as "chromosome puffing." As long ago as 1952, the German biologist W. Beermann suggested that these chromosome puffs were the site of developmental gene activity.[10] The puffs can be so dramatic they increase the bead width fourfold, in the process accumulating high levels of RNA and protein, suggesting that the puffed-up genes are in the process of being expressed. Subsequently new techniques in molecular chemistry confirmed that the puffs did indeed represent gene transcription to messenger RNA, the molecule that carries the nuclear coding of DNA from the nucleus to the protein-manufacturing factories, called ribosomes, in the cytoplasm of the cell. Today, we know that the DNA of genes is condensed within chromosomes by a complex process of packaging around small circular proteins called histones. For a gene to become active, its binding to the histones needs to be loosened up. Puffing is the visible manifestation of this.

In the fruit fly, toward the end of the third larval stage and just prior to pupation, a new set of six chromosome puffs appears. These are known as the "early puffs," and they regress after three or four hours. They are followed by a set of over a hundred "late puffs," which also regress in a very specific time sequence. The cycles of puff appearance and regression on various chromosomes are so regular that scientists could correlate their situation and timing with specific aspects of the metamorphosis of the pupa and the subsequent metamorphosis of the adult insect, for example, with the synthesis of an important protein or with the timing of the eversion of the head from the thorax during the adult metamorphosis. This was the first clue that development—during metamorphosis or during embryonic development within the egg—depends on the activation of specific genes in a tightly controlled time sequence.

In the early 1960s, biologists showed that they could induce typical puffing of chromosomes by injecting purified ecdysone into the bloodstream of insect larvae at a stage prior to metamorphosis to the pupae.[11] Others noticed that, when they injected ecdysone into the fourth-stage larva of the water midge, two puffs were formed within 30–60 minutes, another two appeared at 5–20 hours and two more at 48–72 hours. The location and timing of these puffs was precise and reproducible. This predictability was subsequently confirmed in fruit flies and other insects. It was the first clear evidence that the hormones discovered by Wigglesworth, Williams, and others, and the molting hormone ecdysone in particular, worked by initiating a precise genetic pattern that led to sequential cascades of gene activation during metamorphosis.[12] Now that we know that ecdysones are steroid hormones, in other words, members of a group of endocrine regulators that are found throughout all of life, this is a clue to a more general application beyond the metamorphosis of insects.

The active form of ecdysone, known as 20-hydroxyecdysone, cannot attach to the DNA that makes up a gene. Activated ecdysone must bind to a "receptor" on the chromosome close to the gene that is to be influenced. Today we know that important developmental genes have a number of control sites like this that are located conveniently near to them, sites that can activate, repress, or otherwise influence the expression of the developmental gene. The ecdysone control site is referred to as the ecdysone receptor, and metamorphosis is triggered by the arrival of the hormone at this particular receptor in the insect chromosomes.[13] In some cases, ecdysone causes the early puffs to regress—a repressive action. In others it induces rapid puffing—gene activation. Late puffs are often resistant to ecdysone, indicating they are preferentially responsive to the cascade of new developmental activators and repressors "downstream" of the initial action of ecdysone. Many, though not all, of the genes activated by ecdysone now trigger a complex cascade of activation of some genes and inactivation of other genes, in this way implementing a precisely timed program that leads to the destruction of certain tissues and organs specific to the larva and to the formation of the tissues and organs specific to the adult.

How fascinating, then, that the ecdysone receptors in the chromosomes of insects are almost identical in structure to the thyroid hormone receptors in chordates and vertebrates! This is an important clue to understanding vertebrate evolution and development. Over recent decades, molecular biology has extended our understanding of these developmental pathways, so often initiated by hormones, to show how, at the level of genetic action, the downstream ramifications are formidably complex. It seems likely that similar triggers, systems of response and subsequent developmental mechanisms, control metamorphosis in marine invertebrates. It rather confirms that Wigglesworth was right when, at the very beginning of his career, he suggested that understanding of insect development and physiology would assist our understanding of the development and physiology of all life.

Although there are more than 750,000 species of insects—and some believe many more—they constitute a single class within the phylum of the mandibulates, or arthropods equipped with jaws. Perhaps it is not surprising that metamorphosis in insects has commonalities throughout the entire class. Marine invertebrates, by contrast, are far more diverse, with body forms often dating back to the Cambrian explosion, when all but a single animal phylum first appear in the fossil record, and their evolutionary range extends to many different phyla. This may explain why there is more diversity in the evolution of metamorphic controls and developments in marine invertebrates than we see in insects. In marine invertebrates, the embryonic phase can be very short, and the larva often emerges at a very early stage of development—for example, in the sea urchins, where the egg hatches as a blastula. This develops to a much simpler larva than we see in insects: for example, the wheel-like trochophore of annelids and mollusks or the easel-shaped plutei of sea urchins. In marine groups, the adult is usually much longer living than the larva, with the larval phases usually adapted to dispersal; meanwhile, in many fully metamorphosing insects, such as moths and butterflies, the very opposite applies, with an extended larval phase devoted to rapid feeding and growth, and the adult phase, often brief and sometimes entirely non-feeding, devoted largely to reproduction. All of this would lead us to anticipate fundamental differences between metamorphosis in insects and marine invertebrates.

Insects do have a distant common origin with the marine arthropods, however. And one of the surprising, if fascinating, revelations of developmental biology is that developmental pathways have been remarkably conserved over vast stretches of evolutionary time. This might result in commonalities in the evolution of insects and the related marine arthropod groups—and, perhaps even more widely, all marine groups. One obvious place to look is to the catastrophic metamorphoses that are found in both insects and the marine groups. Are there commonalities in the evolution and development of the imaginal disks of the fully metamorphosing insects and the pluripotent cells that line the coelomic sacs of marine invertebrates?

There appears to be a striking similarity between these "set-aside" imaginal disks and the stem cells that give rise to the radically different adult development in many marine life-forms. The unanswered question, therefore, is how such pluripotent cells, programmed for very different developmental patterns from the larva, might have evolved. Did they arise through mutations affecting developmental pathways? If so, this would imply mutations affecting development at the very earliest stages of embryogenesis, before the dividing cells have acquired any degree of differentiation. Or did they arise, even in part, through hybridization, as Williamson believes? Both explanations involve a certain leap of faith, since they require that the genome has evolved ways of creating separate developmental blueprints within the same fertilized egg and of expressing these separate developmental pathways in strict sequence during the metamorphic process, including the extreme examples, such as the sea squirt, where we witness the two blueprints developing simultaneously. And whichever mechanism one favors, whether mutation, hybridization, or perhaps a combination of several mechanisms, this would still require the evolution of some mechanism for the setting aside of the pluripotent cells, so they remain impervious to the developmental commands throughout the larval stages and only activate when it comes to the development of the adult animal. The origin of set-aside cells is an ongoing puzzle for evo-devo experts, and it may possibly involve an epigenetic contribution.

Minor changes in developmental pathways can have major effects. But once a core developmental pathway has evolved, its very usefulness would appear to confer a resistance to change. This has been elegantly displayed in studies of Hox gene clusters, which play a major role in developments along the longitudinal axis of bilaterian animals, and which show amazing conservation over vast periods of evolution. Moreover, in the evolution of hormonal signaling, some degree of parsimony over the long-term evolution of development also seems likely. In the preface of a recent book on metamorphosis, Lawrence I. Gilbert and his colleagues highlight the fact that "remarkably, the phenomenon of metamorphosis continues to be an excellent model for those interested in the control of development and endocrinological processes."[14] And although amphibian metamorphosis looks very different from insect metamorphosis, here too we find evolutionary conservation of several mechanisms of hormonal control of development: for example, receptors for ecdysones and thyroid hormones are remarkably similar, so much so that artificially reproduced chimerical forms of the hormones are effective in activating target genes not only in the cells of invertebrates but also in vertebrates.[15]

Our own vertebrate ancestors are presumed to have evolved through an amphibian stage. This would suggest that we have something to learn about our own evolutionary development from looking at what happens in amphibians, many of which undergo striking metamorphosis.

— 23 —

A Tale in a Tail

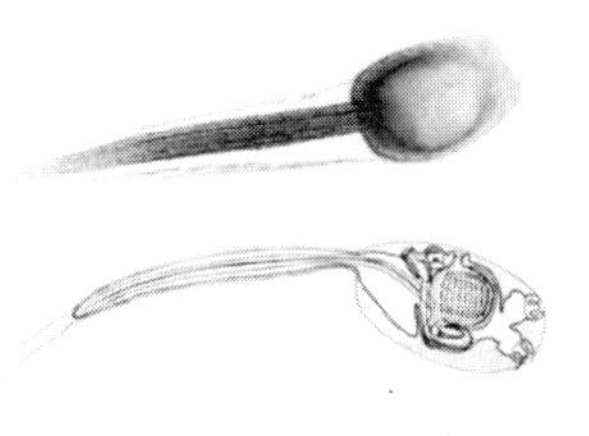

Tadpole larva of sea squirt

Vertebrate evolution, with its chordate ancestry, has a surprisingly ancient history, and some biologists believe that the sea squirts, with their tadpole larvae, may represent the earliest known stage of that history. The sea squirts, or ascidians, belong to what is variously classed as the phylum, or subphylum, of the urochordates—essentially headless creatures with the rudiment of a spinal cord: frogs and toads, by contrast, belong to the phylum of the craniates, the more familiar chordates that do have heads. And while sea squirt larvae bear a remarkable resemblance to the tadpoles of frogs and toads, in particular in sharing the chordate locomotory tail, it would be disingenuous to equate the two. The ascidian larva is only two or three millimeters long, is incapable of feeding, and possesses an internal structure that is much simpler than that of the amphibian tadpole. Moreover, its evolution vastly pre-dates that of the amphibians. Giant tadpoles have been described from the late Cambrian in China, which would make them contemporaries of the early marine invertebrates, but, as Donald Williamson points out, they could not have been the larvae of amphibians, because the amphibians did not arrive onto the scene for another two hundred million years.

Some evolutionary biologists agree with Walter Garstang, who proposed that evolutionary changes at the embryonic or larval stages were capable of changing the evolution of the adult form. From this perspective, chordates might have originated from ascidian tadpoles through larval features being carried into the adult form—an example of Garstang's mechanism of "paedomorphosis." E. J. W. Barrington, a former professor of zoology at Nottingham University, agreed with Garstang's theory, suggesting that one possible way in which tadpole forms might have persisted into adulthood without metamorphosing was through failing to respond to environmental cues.[1] For example, a tadpole larva that normally metamorphosed on contact with algae tethered to the ocean floor might have drifted or swum into deep water and simply failed to discover the metamorphic signal. But for this to lead to paedomorphosis we would have to assume complex additional evolutionary developments, including the ability of the tadpole stage to reproduce. It might also have needed to feed—though the inability of the present-day tadpole to feed could be a later adaptation.[2]

According to Williamson, this complex rigmarole to arrive at chordate evolution is unnecessary. In his view the forerunner of the vertebrates was an adult marine chordate with tadpole features, rather like the Cambrian fossil tadpoles found in China, which hybridized with the ancestral tunicate to give rise to the sea squirt metamorphic life history. Williamson also suspects that hybridization between a marine vertebrate with tadpole features and a non-metamorphosing protoamphibian might have given rise to the amphibian tadpole metamorphosis. While this might appear a bit of a stretch, the truth is that it would be difficult to propose any theory that did not stretch the imagination in order to explain how, in the words of Barrington, "the ascidian larva is a dual organism in which there is sharp demarcation between the temporary organs of the larva and the rudiments of the permanent organs of the adult." The closer one examines this strange duality, the more bizarre it seems. Inside the fertilized sea squirt egg, the embryo develops, in conventional fashion, to the stage of a gastrula. "From this point onward," in Barrington's words, "two independent developmental mechanisms are operating side-by-side, the development of the larval structures proceeding virtually independently of that of the permanent ascidian organization."[3] Whatever the evolutionary origins for this remarkable situation, it must explain how the fertilized egg of an ascidian contains not just a single distinct developmental program but two, as well as how these two programs, once past the common gastrula, are capable of developing simultaneously into two radically different life-forms.

Presumably, very early in embryology, two stem cell populations must separate out—one is reminded of the origins of identical twins in humans—but in this case each stem cell population is programmed for independent developments, with the tadpoles and juveniles growing side by side. Initially the tadpole development proceeds a little faster, but the juvenile soon catches up. Meanwhile the very different larval and juvenile nervous systems also develop independently, whether side by side, or as one above the other, until eventually the tadpole settles to the ocean floor, its body degenerating, leaving behind the juvenile adult. Moreover, though both the adult and the larva are bilaterian, the developing juvenile follows a completely different body orientation from that of the larva. In the conventional evolutionary trajectory, one might expect the adult brain to develop, through adaptation, from the larval ganglion, but in fact the adult develops an entirely new brain, and the tadpole ganglion is discarded.[4]

The presumed evolutionary relationship between these headless chordates, the ascidians, and the vertebrates suggests that we might look to the simpler forms to help us understand the evolution and developmental pathways of more complex life-forms. But as long ago as 1968, Barrington urged that we do so with caution.[5] Sea squirt metamorphosis takes place in relatively simple animals, and it is initiated at a very early stage in their development. The regulation of their development, including the dramatic metamorphosis, may not require any more complex mechanisms than would be found in the early embryonic development of a more complex animal. Even so, it is nevertheless interesting, in the light of what we have learned from the hormonal triggers of insect metamorphosis, to look for commonalities in the metamorphosis of marine invertebrates and that of chordates, from ascidian tadpoles to vertebrates. This perspective assumes, however much it strains credulity, that we humans are distantly related to sea squirts, as the notochord and its accompanying nerve cord in the tadpole larvae, or tadpole adults, suggests. There is also another, more subtle thread of causality that links the metamorphosis of the ascidians to chordate evolution, one that is still detectable today in our human brain.

We have already seen that marine invertebrate metamorphosis is primed to respond to environmental cues. D. Jackson and his colleagues at the University of Queensland have investigated the metamorphosis of the sea squirt, *Herdmania curvata*, which lives on the underside of boulders and ledges on coral reefs, observing the dramatic transformation that begins when the appropriate environmental cues come into contact with tiny sensory organs, known as papillae, on the skin over the head of the tadpole larva.[6] These papillae send a signal along sensory nerves to local cells that release a key internal factor, and this in turn triggers the developmental programs that control metamorphosis. Other biologists have demonstrated that this internal factor is probably a protein, known as "Hemps," which is localized to those same sensory papillae on the tadpole's head. Much as we have seen with ecdysone in insects, Hemps plays an important role in regulating developmental pathways, triggering a developmental cascade involving hundreds of genes, which program the destruction of the tail and other unwanted larval organs and tissues, meanwhile simultaneously masterminding the development of the juvenile adult organs, tissues, and body form. All of this happens with amazing rapidity. Within minutes of the larva coming into contact with the environmental cue, retraction of the head papillae and resorption of the tail are seen to begin.[7] And here a more subtle developmental commonality with more advanced vertebrates comes into play.

In sea squirts, an organ known as the endostyle lies buried in the floor of the throat. This connects to the gut through a duct along which a mucus secretion is passed from the endostyle. The role of the endostyle is to fix iodine—in other words it takes up free iodine from the oceanic environment and chemically binds this to organic molecules. This process, known as iodination, is a necessary step in forming thyroid hormones. Recent studies by Professor E. Patricolo and colleagues in Italy have confirmed the presence of the thyroid hormone thyroxine in the tissues of sea squirt tadpole larvae.[8] Thyroid hormone also appears to play a key role in ascidian metamorphosis, so much so that when the same researchers administered a poison known to disrupt thyroid function, the larvae failed to metamorphose to the adult form.[9] Other biologists have taken this a step further. The developmental gene *CiNR1* is linked to thyroxine activation. In 1998, E. Carosa and colleagues demonstrated that *CiNR1* was active in the sea squirt *Ciona intestinalis*.[10] In another primitive chordate, known as the lancelet, metamorphosis takes place through a larval stage, and its larva has an endostyle where iodine is fixed and where thyroid hormones have been discovered. These findings have led biologists to surmise that the endostyle may be the evolutionary forerunner of the vertebrate thyroid gland.

Thyroxine has two very different biological actions, one developmental and the other endocrine. The difference lies not in permutations of the hormone but in the place and timing of its action, whether during development or in adult physiology. Thyroxine acting hormonally in adult tissues works as a control lever on the rate of cellular chemistry. Thyroxine acting on embryonic or metamorphic tissues acts as a developmental master switch. When it enters the nucleus, it discovers and binds to special receptors within the chromosomes, and this triggers key control genes, which turn on developmental programs. This pattern of regulation is vital to the way in which embryonic development is controlled. It is also vital to the genetic cascades that underpin metamorphosis. From the evolutionary point of view, it is interesting that the same subfamily of nuclear receptors, known as the thyroid receptor, or TR, responds not only to thyroid hormones but also the molting hormone, ecdysone.

Thyroxine has been shown to play an important role in the development of sea urchins, crown-of-thorns starfish, and the sand dollar, *Leodia sexiesperforata*.[11] This suggests that development can be so conservative that we are going to find interesting evolutionary links between widely different phyla.

Lampreys are primitive cartilaginous fish, some of which are free-feeding while others are parasitic on other fish. They are jawless and easily recognized from the sucker that surrounds their mouths and by the single nostril on the top of their heads. Their skin is devoid of scales and slimy to the touch, and their seven gill openings extend backward from the line of their eyes. The parasitic varieties attach to other fish by means of the sucker, using their teeth-bearing tongues to rasp through the scales and suck their blood. Others feed on small invertebrates. All lampreys have a cartilaginous skeleton that does not mineralize to bone. Even the marine varieties return to freshwater rivers and streams when it is time to spawn. After hatching, lampreys spend most of their lives in a larval phase, known as the ammocoete. This lasts from three to seven years, during which time they burrow in the soft sediments of stream and river bottoms and filter-feed on detritus. The metamorphosis is relatively slow, taking three or four months, during which time the animal undergoes extensive internal and external alteration, including the development of the eye, the loss of larval bile ducts and gall bladder, the transformation of the lining cells of the intestine and gills, the total regression of the larval kidney and its supplanting with the adult kidney, and the development of teeth and tongue suitable for the adult life cycle. During the metamorphosis, the larval endostyle develops into a fully operational thyroid gland, with secretory follicles. The marine varieties of lampreys then return to the sea, where they spend a year or two in the adult phase of the life cycle before returning to the rivers to spawn, and usually to die.

The lamprey, though highly adapted to its modern lifestyle, is still widely regarded as a living representative of the early evolution of the vertebrates, so this transformation of larval endostyle to a thyroid gland during metamorphosis is of particular interest to developmental biologists. In the early decades of the twentieth century, scientists attempted to trigger lamprey metamorphosis with thyroid extract. But all such efforts failed, suggesting that thyroid hormones played no part in lamprey metamorphosis. In fact this interpretation was quite wrong—but for an interesting reason. In lampreys thyroid hormone acts like juvenile hormone in insects, inhibiting rather than stimulating metamorphosis.[12] In bony fish, such as eels and flounders, the situation is the exact opposite, so that treatment with thyroid hormones induces precocious metamorphosis. As long ago as 1928, E. Murr and A. Sklower studied six arbitrary stages in the metamorphosis of the eel, during which they found that the volume of the thyroid gland increased some fourteenfold during the early stages,[13] and today many studies have confirmed the role of thyroxine and its pituitary-driven negative-feedback control, through thyroid stimulating hormone, in the metamorphosis of a wide variety of bony fish, including flounder, eels, flatfish, zebra fish, sea bream, coral trout grouper, and the orange-spotted grouper. These observations may have commercial applications for breeding in fish farms. More importantly, they have evolutionary considerations. Fish are a vital stage in vertebrate evolution. At some time in the distant past, a fishlike ancestor emerged onto land to give rise to the first four-legged animals, or tetrapods, which in turn evolved to the terrestrial vertebrates. The next stage in this complex, many-branching, and sometimes reuniting, evolutionary trail, are the amphibians, which also share the mystery of metamorphosis.

— 24 —

Of Frogs and Their Relatives

African clawed frog

The life cycle of frogs and toads, with their metamorphosis through a tadpole stage, is every bit as enchanting to schoolchildren as the quasi-magical transformation of caterpillars to butterflies. In fact, the closer we look at these familiar amphibians, so similar in form to ourselves that they have become the stuff of fairytales, the more interesting their real life story becomes. They really are amazing creatures.

The first surprise, when we examine the evolutionary origins of frogs and toads, is to find that they are so exceedingly ancient. Amphibians are thought to have evolved from a group of lobe-finned fish known as rhipidistians. These gulped air and dragged themselves through the mud at the water's edge using their bony fins. In time the fins evolved to the four limbs of the first land-living vertebrate animals. Spectacular fossils of this evolutionary step from the oceans to land were discovered in Canada, showing a fish-like animal called *Tiktaalik*, which was between four and ten feet long and dates to approximately 375 million years ago. It appears to be intermediate between fish and reptiles, with a typical fishy body, jaws, fins, and scales, but with the head, neck, and ribs of a land animal. Within the fins, the bones have evolved to an intermediate stage, between swimming and terrestrial weight-bearing.[1] But these pioneering vertebrates did not necessarily share our familiar five digits on the ends of their four evolving limbs. Three different genera of early animals have now been discovered, *Ichthyostega*, *Acanthostega*, and *Tulerpeton*, all dating to the late Devonian period some 390 to 340 million years ago, with, respectively, seven, eight, and six toes.[2] The oldest reptile fossils are almost as ancient as those of the amphibians, suggesting that the amphibians and the early reptiles, birds, and ancestral mammals diverged very early from a common ancestor, perhaps through differing adaptations to the amphibian and terrestrial life cycles. The earliest frog fossils appear during the early Jurassic period of North and South America, about 208 million years ago, which makes them contemporaries of the early dinosaurs. But where the dinosaurs, other than those that evolved into birds, became extinct, frogs flourished.

Today's frogs have diversified to more than four thousand estimated species, and they have extended the range of ecosystems they inhabit to every continent other than Antarctica. By any measure, this is an extraordinary evolutionary success story. It simply begs an explanation.

Of course frogs are familiar as small, tailless animals with a squat body and long, powerful hind legs adapted to jumping. They have large bulging eyes and moist skin. Typically they live on land or in trees close to water, and they come equipped with lungs that breathe air when out of the water. Surprisingly, they can also breathe through their skin when underwater, because their skin is richly supplied with blood vessels so oxygen can diffuse through it and enter the blood. Toads, by comparison, have a dry warty skin and prefer to live in drier places. In spite of their ancient colonization of land, the great majority of frogs lay eggs in or around water so that their larvae, familiar as tadpoles, hatch out in this ancestral habitat. Amphibian tadpoles have no limbs, breathe through fishlike gills, and swim with a powerful fishlike tail. This dual existence of two developmental forms in one life cycle, adapted first to water and then to land, is characteristic of the amphibians, a class that also includes toads, newts, and salamanders. It distinguishes amphibians from the class of reptiles, which includes turtles, snakes, crocodiles, and alligators, not to mention the now extinct dinosaurs.

The four thousand different species of frogs range from a Brazilian example a third of an inch long, which weighs less than an ounce, to a West African goliath, which measures a foot and tops the scales at seven pounds. Frogs also have a jumping locomotion that enables them to leap up to twenty times their body length, a remarkable adaptation that puts our Olympic long jumpers into the shade. Like the capacity to live, and breathe, in water and on land, one can only imagine that this impressive adaptation must have helped them escape in times of danger. To accommodate the leap, frogs have evolved very long legs in relation to their body size and a specialized pelvic girdle designed for power and shock absorption.

It will come as no surprise to learn that there are many variations of development among the amphibians, and within the large and diverse group of the frogs and toads, there are exceptions to the metamorphic life cycle. In some species the female gives birth to tiny, fully formed frogs, the tadpole phase subsumed within the mother prior to birth. Nevertheless, the familiar metamorphosis from tadpole to frog is typical. "A remarkable series of changes takes place when an aquatic fishlike tadpole is transformed into a land-dwelling frog," in the words of James Norman Dent, based at the Department of Biology at the University of Virginia, Charlottesville.[3] This includes massive destruction of larval tissues and organs and a major reconstruction of the new adult. Some of these changes arise slowly and progressively within the lifetime of the tadpole, while others take place very rapidly, while still in the water, at the time of metamorphosis. The latter begins with the formation of a membrane, called the "operculum," that grows over and covers the gills of the tadpole. The head, meanwhile, is extensively remodeled as the larval cartilage is remolded to the adult bony cranium, and the horny frog beak begins to form. New color patterns invade the skin. The front and back legs, which had already begun to develop during larval life, rapidly elongate, and the tail is reabsorbed, causing the anus to move to a different position. Inside the body, there are dramatic changes to the nervous system and eyes, wholesale reconstruction of the great arteries from the heart, shortening of the gut, degeneration of the larval skin to be replaced by the adult skin, and formation from scratch of the hyoid cartilages in the throat, which, together with growth of the chest muscles, make it possible for the frog to breathe air.

At a molecular level, amphibian metamorphosis is accompanied by the reprogramming of genes expressed in the amphibian liver, the switching of genes responsible for types of hemoglobin, changes in the expression of the gene responsible for the hardening of skin, and dramatic changes in the immune system.[4] As is now familiar with all varieties of metamorphosis, these genetic cascades bring about the massive destruction of larval tissues and organs, with death and dissolution of unwanted cells, together with wholesale reprogramming and regeneration at every level from genes, through cells to tissues and organs.[5] Inevitably scientists have searched for a role of the thyroid gland in these catastrophic algorithms of destruction and reconstruction.

Thyroid extract was found to hasten frog metamorphosis as early as 1912. Over subsequent decades, it became clear that thyroid hormone is fundamental to the entire process, activating key control genes that initiate the genetic cascades of metamorphosis. And just like ecdysone in insects, the production of thyroid hormone is subject to negative feedback in response to a master gland in the head, the pituitary. This feedback system kicks in during the early larval stages of the amphibian life cycle.[6] We humans share the same pituitary-thyroid axis. Like frogs, our pituitary gland is linked to a part of the midbrain known as the hypothalamus, which is exquisitely responsive to external cues.

In response to some priming external cue, which is conveyed to the hypothalamus, the developing amphibian pituitary gland secretes thyroid-stimulating hormone, TSH, and this is carried through the bloodstream to the thyroid gland, where it results in the release of thyroid hormone. This, in turn, not only triggers metamorphosis but also, through negative feedback, switches off the pituitary release of further TSH. The body structures, internal organs, and biochemistry of amphibians are a good deal more complex than those of insects and, for the complete panoply of metamorphosis to take place, other hormones are also important, including prolactin, adrenal steroid hormones, and melatonin, a hormone that controls skin pigmentation and responds to prevailing light and dark, possibly cued by the duration and intensity of light falling onto the eye. The duration and cycles of daylight versus night have been shown to influence the speed of metamorphosis, with observable effects not only on the rate of change but even the rhythms of cell proliferation in skin. According to Jane C. Kaltenbach of Mount Holyoke College, "it is possible that the effects of lighting changes . . . are mediated by the pineal gland and [the hormone] melatonin and are partially responsible for the 'occurrence' of spontaneous metamorphosis when conditions in nature are favorable." The pineal gland is adjacent to, and intimately connected with, the pituitary, which would allow environmental cues to trigger hormonal secretion. Thus, for example, cues arising from low environmental temperatures also slow down, and may even inhibit, metamorphosis.

Donald D. Brown, an adjunct professor in the Department of Embryology at the Carnegie Institution of Washington, D.C. For many years he and his colleagues have been researching the way thyroid hormone controls metamorphosis in amphibians. In particular they have studied metamorphosis in the African clawed frog, *Xenopus laevis*, a plump medium-sized aquatic frog with smooth, slippery skin, webbed rear feet, and clawed front feet.[7] In a series of experiments, they have introduced mutant genes into the sperm of the frog to create an abnormal thyroid hormone receptor protein in the offspring. This competes with the normal receptor during metamorphosis, when the mutant receptor blocks the normal developmental response to thyroxine. The result is a catastrophic failure of every metamorphic change in the tadpoles. From such experiments, Brown has concluded that metamorphosis in frogs and toads follows a set of complex developmental programs in which the entire process is controlled by thyroid hormone. We have already seen how, during amphibian metamorphosis, tissues and organs grow de novo, die, or are remodeled. The experiments of Brown and other researchers have shown that thyroxine works by changing the expression of genes in these tissues that are remodeled or die during the process of metamorphosis. Moreover it has been shown that thyroxine achieves this effect in two ways, directly at single-cell level, where it controls cell differentiation into specific types necessary for tissue and organ development, and also at the level of cell-to-cell signaling within tissues and between adjacent tissues. This is another important step in understanding not merely amphibian metamorphosis but also development throughout all of multicellular life.

Vincent B. Wigglesworth did not live long enough to witness the first unfolding of the human genome. I have no doubt that he would have found it fascinating. For me, as for many biologists, one of the most revealing aspects was how much genetic programming we have in common with every other form of life on Earth. For example, we share 2,758 of our genes with the fruit fly and 2,031 with the nematode worm; and all three of us—human, fly, and worm—have 1,523 genes in common. Even more pertinent is the common evolution of Hox genes, which help to regulate embryonic and post-embryonic development along the front-to-back axis of all animals.

So important are the Hox genes that they are now seen as "a guiding force within the field of evolutionary developmental biology."[8] We have already seen how ancient these Hox genes really are and how remarkable is their conservation from marine invertebrates to flies and humans. To put it in the words of D. E. K.Ferrier and C. Minguillón, "The finding that galvanized the EvoDevo community was that in the Hox genes we have [similar] genes acting in [much the same way] across the animal kingdom."[9]

Common genes, common developmental control mechanisms—perhaps it is not altogether surprising that key commonalities should extend from the developmental mechanics of metamorphosis in marine invertebrates, through primitive chordates, through marine vertebrates, through the common ancestor of amphibians, reptiles, and mammals, to humans.

— 25 —

The Wonder of Development

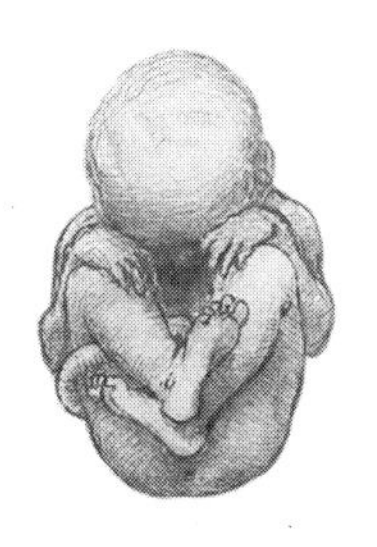

Human baby in the womb

"Man," to quote the words of Jacob Bronowski, "is a singular creature. He has a set of gifts which make him unique among the animals: so that, unlike them, he is not a figure in the landscape—he is a shaper of the landscape."[1] What makes us quintessentially human is not the fact we can walk on two legs—birds do—or the fact we have exquisitely dexterous limbs—octopuses have eight of these. We are human because we, alone, have the gift of advanced sentient intelligence. How then can we come to understand how our singular species belongs to the same natural world as does a frog or a honeybee, and yet from out of the complexity of nature's history we have become, in Bronowski's words, "explorers of that nature"?

Early in the evolutionary divergence of terrestrial vertebrates, the lines leading to the amphibians split off from a common ancestral line that led first to the reptiles and later to the mammals. At this key point of divergence, the metamorphic inheritance underwent a fundamental schism. The amphibians, in bridging the two very different ecologies of water and land, evolved a new developmental drama of metamorphosis; meanwhile the common reptile-mammalian line abandoned it.

The lengthy development of the fetus within the mammalian womb allows more time for complex embryonic development than was possible for our egg-laying ancestors, and this has subsumed many traces of any metamorphic inheritance. But, though we cannot see it from the outside, the quasi miracle of hormone-controlled development is still there. The pregnant mother can sense some of the changes taking place inside her. She feels it directly when fetal hormones affect her body's chemistry, and she thrills to the first tiny movements of the developing fetus. And, looking back over our shoulders at the mammals as a class, here and there we glimpse reminders of that dramatic metamorphic past.

Metamorphosis is now viewed as a new phase of "post-embryonic development" and, as we have seen in chapter 13, we humans exhibit a singularly important example of post-embryonic development at the time of puberty, suggesting to some biologists that puberty in humans is a form of metamorphosis.[2] Much as metamorphosis in insects and amphibians is brought about by signals that begin in the central nervous system, leading to the release of ecdysone and thyroid hormone, puberty in humans is also brought about by pulses of gonadotropin-releasing hormone, which is released by the hypothalamic portion of the brain. This stimulates the neighboring pituitary gland to increase the secretion and release of the sex-gland-stimulating hormones, or gonadotropins, which travel through the bloodstream to the ovaries and testes, where they stimulate increasing blood levels of estrogens and androgens. Just as environmental cues can trigger metamorphosis, environmental factors can also influence the timing of puberty; for example, better nutrition may have lowered the age of female puberty during the last century, and environmental contamination with bisphenol A, a common component in plastics production, may also hasten the onset of puberty.[3] Puberty, and human fertility, can also be considerably influenced by physical illness and environmental stress, the latter almost certainly mediated by input through the hypothalamic link,[4] echoing the environmental signaling we saw in amphibian reproduction.

Moreover, puberty is not the only time of post-embryonic development in humans. Unlike most other animals, humans retain the embryonic potential for brain development long after birth, a facility that allows our exceptionally large and complex brains to grow and develop throughout the first two years of infant life.[5]

The human body contains about two hundred different types of cell, specially adapted for the different tissues and organs. Almost all of these different cells, tissues, and organs are formed during embryonic development. As T. Mohun and J. Smith explain, we can observe a similar specialization happen in frog spawn with little more than the assistance of a magnifying glass.[6] The embryo changes rapidly and accurately from the ball of cells known as the blastula to more complex shapes containing the first evidence of distinct tissues, from brains to limbs. Research on how such developing cells recognize their own tissue type is the basis of the present wave of stem cell research that may soon help doctors treat serious illnesses, such as Parkinson's disease and the damaging effects of heart attacks.

Looked at more broadly, it is clear that the entire developmental plan of a human being must be contained within the egg immediately after fertilization. But this does not mean that every cell derives from an autocratic central organization, with no room for variation. We see this potential for "downstream changes" in the slight differences between identical twins, which is brought about by epigenetic control over gene activation as opposed to genetic variation—an astonishing additional layer of environmentally influenceable control over our genes, both within the womb and throughout our lives from infancy to death.[7] Indeed those same epigenetic control systems are critical in deciding the fate of embryonic stem cells, whether they will become liver, blood, or brain. So it is that with modern advances in developmental biology, we realize that embryonic development is determined by a combination of mechanisms: the central "genome-determined" plan, the critically important epigenetic systems of gene control, and an additional complex series of interactions between adjacent cells and tissues.

All of these potentials are preprogrammed into the human egg from the moment of its fertilization.

The idea that every individual human being arises from a single pluripotent cell never ceases to amaze me. We have evolved from humble beginnings, much as any other life-form, and we develop from a fertilized egg along similar developmental principles to other animals. And like all multicellular animals, plants, fungi, and many of the single- or few-celled life-forms known as protists, we are composite beings. We breathe oxygen thanks to the mitochondria that originated at least a billion years ago from the symbiotic incorporation into our ancestral cells of free-living, oxygen-breathing bacteria. A significant part of our genetic inheritance came into being from the progressive incorporation of large numbers of retroviruses, similar in genetic structure to HIV-1, the cause of AIDS, which, like the bacteria that became mitochondria, have long since been incorporated into our normal genetics and development.[8] At least eight of these endogenous retroviruses appear to be playing important roles in placentation and normal human reproduction.[9] We are dependent on symbiotic bacteria in our guts for healthy digestion and on myriad plants and other animals for the vitamins, essential fatty acids, and essential amino acids to keep us healthy from day to day. And still, the connections that integrate us into the biosphere of our beautiful ocean-dominated planet extend deeper, to the most profound levels of our biochemistry, our physiology, the genes and chromosomes that are responsible for our programming and inheritance, and for the cellular structure and machinery of development that enables that wonder of every individual human life.

On Day 1 of your origins, the first cleavage of that fertilized cell took place. You were now two cloned cells, each inheriting matching pairs of 22 chromosomes from each of your parents in addition to the two sex-determining chromosomes, making up the human chromosome complement of 46. If, in addition to the single X chromosome from your mother, you inherited an X chromosome from your father, you are now female; if a Y chromosome, you are now male. By Day 4, you had become 16, then 32 cells, a solid planula-like ball known as a morula. The following day the morula developed into the familiar hollow embryo known as a blastula, with an outer wall of cells and an inner cell mass surrounding the fluid-filled core. You subsequently developed from the inner cell mass, known as the "embryonic disk." Over the following couple of days your blastula, now known as a "blastocyst," attached to the lining of your mother's womb, where it implanted itself, making itself a new home. At the area of contact with the womb, your endogenous retroviruses controlled the cellular differentiation of the placental interface, fusing the cells into a confluent monolayer—at the same time helping to make the human placenta the deepest, most invasive, and highly specialized reproductive interface of all the mammals, shared, along with the same viruses, only with the great apes. Your blastocyst began to secrete the hormone human chorionic gonadotropin, or HCG—the chemical basis of the popular pregnancy tests—which entered your mother's circulation and stimulated her ovaries to produce estrogen and progesterone to prevent menstruation.

It did not trouble the emerging you that it also made her liable to morning sickness. Over days 7 to 10, the indentation known as a "blastopore" pushed into your blastula wall, signaling that you were turning into a "gastrula." Up to this point, each of your cells was still a clone of that first fertilized cell. But with gastrulation you underwent the first developmental change that would differentiate your cells into the three fundamental tissues that would fashion all of your adult tissues and organs: the "ectoderm" that would become your skin and nervous system, the "endoderm" that would form the lining of your gut and your internal organs, and the "mesoderm" that would become your muscles, bones, and heart. Your blastopore would not become your mouth, but the opening at the latter end of your intestinal tract, your anus. Your future mouth would have to break through to the surface ectoderm at the other end, making you, like all other members of the chordate lineage, a deuterostome. The shared inheritance runs even deeper. Your abdominal lining, or coelom, would develop from outpouchings of your primitive gut, making you an "enterocoelous deuterostome."

By days 15 to 21, your developmental inheritance, and in particular your very ancient developmental genes, your Hox genes—quadruplicated as a result of two rounds of whole genome duplication, probably the result of two major hybridization events long ago[10]—began to control development along the length of your body. Your body form assumed a typical bilaterian shape, with the front end taking on the first indications of a head and the rear end taking the form of a tail, with two parallel rills appearing along your front-to-back axis, heralding the "neural groove" that would fold over into a cylinder to form your future spinal cord and, at the head end, the brain that is currently directing the reading of this book. By Day 21 you had vestigial gills, Haeckelian relics of a marine past, together with a more pronounced stubby tail. From weeks 4 to 8, all the major organs of your body were formed. As early as 12 weeks most of your organs and body parts were developed, with the exception of your brain and lungs. From this point onward, the bulk of your organs and tissues were now developed, though they still needed to grow, but your brain was still largely undeveloped.

The extraordinary thing, yet a thing that is absolutely obvious when we think about it, is that when you first arrived into the world, as a newborn baby, your brain was still relatively small—little more than a quarter of its present size—and much of its cortical, or higher, functions were lacking in development.

During the early days after birth, your brain added approximately a quarter of a million nerve cells every minute to its mass, and this extraordinary increase in brain size and complexity continued for the first two years after your birth. If we compare this to our closest evolutionary relatives, the chimpanzees, the ratio of brain weight to total body weight is the same in both species at birth. But such is the post-embryonic growth of the human brain that by the time we are adults, our brain to body weight is three and a half times that of the chimpanzee.[11] The brain works through the connections, known as synapses, between different nerve cells. At the cellular level, roughly 30,000 synapses per second per square centimeter of brain cortex are formed during the first few years of life.[12] Indeed, it is this extraordinary development of our brain, during both our embryonic and post-embryonic stages, that makes us quintessentially human. And here, also, we discover a telling link to the metamorphosis story.

26
The Human Brain

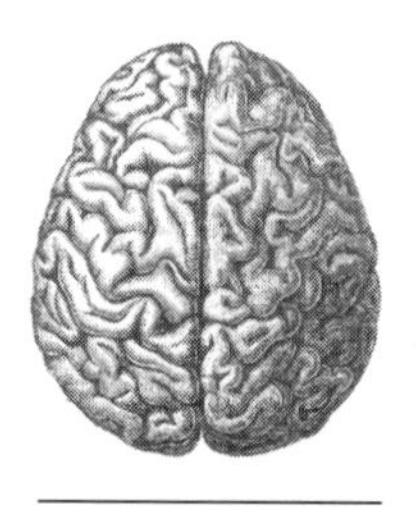

Human brain

In 2004, in an editorial published by the Endocrine Society, R. Thomas Zoeller of the Biology Department at the University of Massachusetts Amherst made the intriguing statement that "insights gained from the studies of frog metamorphosis are helping us understand the role of thyroid hormone in the development of a completely different tissue—the human brain."[1]

Zoeller did not start out his academic life with an interest in developmental biology. After a spell in the army, from 1972 to 1974, he spent some time with a fish and wildlife manager, which in turn inspired an interest in biology and ecology. He decided he would enroll as an undergraduate at Indiana University. Here a group of scientists, including Robert W. Briggs, were studying amphibian development, using the frog *Xenopus laevis*. As a graduate student, Zoeller moved to Oregon State University, where he was inspired by the work and personality of Frank Moore, a young and enthusiastic professor, who was fascinated with the hormonal control of behavior. For his PhD, Zoeller studied the physiological basis of reproductive behavior in the rough-skinned newt, *Taricha granulosa*, and in particular the coordination between environmental factors and the motivation to reproduce.

To put it simply, the rough-skinned newt would only engage in sexual behavior in certain circumstances. At its most obvious, the male newt would only do so in the presence of a sexually competent female. This may sound obvious, but if we take a mental step backward and think a little harder about what is really going on within the newt brain and endocrine system, there must be visual signals to tell him the female is nearby, perhaps pheremonal signals to tell him she is reproductively competent, and very likely a combination of the two. As we have seen in so many examples of metamorphosis, these external cues must activate internal responses, in this case biochemical and very likely hormonal. Zoeller examined minute samples of the newt brain, looking for any chemical responses that would help him understand the way in which this complex behavioral system actually works.

In time he did find some changes in different levels of certain chemicals, known as neuropeptides, which looked like a significant response. He pushed on, becoming interested in a neuropeptide known as arginine-vasotocin. But in the end he was frustrated in arriving at his PhD conclusions because merely looking at changes in peptide content did not tell him enough to really understand the system.

In 1984, having finished his doctorate, he moved to the National Institutes of Health, where he studied the nerve cells that control the hormonal system of reproduction. Entomologists such as Wigglesworth and Williams had shown that an external signal or cue sends a message to key cells in the insect brain, which then initiates the internal and behavioral responses of metamorphosis. In vertebrates, such as frogs and humans, a distinct population of brain cells, known as the gonadotropin-releasing-hormone cells, or GnRH cells, control reproduction. Scientists at NIH had found a way of identifying GnRH within these brain cells, which allowed them to study the regulation of sex hormones in reproduction. The youthful Zoeller heard through the grapevine that Peter Seeburg, then working at Genentech, had cloned the rat gonadotropin-releasing-hormone gene. At much the same time, the chemical structure of thyrotropin-releasing hormone, or TRH, was published. TRH regulates the release of thyroid hormone through control of the pituitary-thyroid hormonal axis.

At NIH, Zoeller began to investigate some of these key nerve cells, or neurons, in the brain. He began working on the cells that make gonadotropin-releasing hormone—GnRH—but he soon ran into a problem. The cells that make GnRH are not gathered together in a distinct cluster, or nucleus, but are scattered throughout a region of the hypothalamus. This made it difficult to work with the GnRH system. Thus in 1985, and wholly for pragmatic reasons, Zoeller shifted his focus to the thyrotropin-release system, where the cell bodies of the nerve cells that make TRH happen to be focused in a single compact collection, known as the paraventricular nucleus. From here their threadlike nerve strands, carrying nerve signals and hormones, travel out of the nucleus to enter the hypothalamus. As we have touched on already, the hypothalamus is one of the most important structures in the vertebrate midbrain, playing a key role in the regulation of the master gland, the pituitary. We might also recall that this is a critical part of the tadpole brain during amphibian metamorphosis. All of the brain-derived tropic hormones, such as the gonadotropin- and thyrotropin-releasing hormones, converge on the hypothalamus, where a system of blood vessels ferries them to the pituitary.

In 1986 Tom Zoeller's son was born. Up to then he had thought only dimly about embryonic development and little or not at all about the role of thyroid hormone in development. In his words, "The birth of a child should change you. If it doesn't change you at the moment of birth, it's gonna change you later." It certainly changed Zoeller's interpretation of what was really important. In 1988 he moved to the University of Missouri School of Medicine, where he worked on two projects: the negative-feedback system of thyroid hormone on neuronal levels of TRH; and the effects of alcohol on this same system. At this time it was known that deficiency of thyroid hormone in the human fetus could lead to severe mental retardation as well as difficulties with movement and coordination in later life. But knowledge of the underlying mechanisms was hampered by a misunderstanding. Many scientists at the time believed that the placenta broke down maternal thyroid hormone, preventing it from entering the fetal tissues, and so doctors assumed that thyroid hormone had no effects on brain development until after birth, when the infant's own thyroid gland became functional. But a number of recent investigations had thrown doubt on this. For example, from 1989 onward, several colleagues had shown that thyroid hormone of maternal origin appeared to cross the placenta and enter the fetus.[2] If thyroxine was important for brain development this made sense, since most of the key developments in the mammalian brain took place before birth. But this introduced a new question.

If maternal thyroid hormone got into the fetal circulation and from there into the developing tissues, what was it doing there?

Only recently other researchers had discovered something rather puzzling: the levels of the activated thyroid hormone, known as triiodothyronine, or T3, in the developing cerebral cortex of the human fetus were much higher than would be expected from the circulating levels of the hormone in the fetal blood.[3] This made Zoeller sit back and think again. "The more I thought about [it] the more I realized that the time when thyroid hormone exerts a very profound effect on the brain must be during [embryonic] development."

In 1992, he read a key paper by Scott Young and his colleagues, in which they probed the fetal brain, searching for developmental genes that might be targeted by thyroid hormone.[4] For a hormone to target a developmental gene, there must be a hormone receptor site in the chromosomes close to the gene. There are two main types of thyroid-hormone receptor in the brain, known as alpha and beta receptors. Young and his colleagues were able to map the genetic expression of these two thyroid receptors (TRs) in mammalian brain cells in terms both of their spatial distribution and their expression over time. Zoeller was fascinated with the results: an exquisitely outlined pattern and complexity of thyroid-receptor expression, which must surely be the key to what thyroid hormone was really doing during fetal development.

Zoeller began to design a new series of experiments to test this further. But it wasn't until he made another move, to his current base at the University of Massachusetts Amherst, where he had recruited a graduate student, Amy Dowling, that he was in a position to conduct the first experiment. If, as he now suspected, thyroid hormone plays a key role in mammalian brain development, specific genes within these developing brain cells must be critically responsive to the presence of thyroid hormone. He needed to find a way of identifying those genes.

He knew that other scientists had looked for the same genes and failed. "I thought the reason they failed was that they were looking at the differences between fetal brain development in a normal pregnant animal and an animal that had been deprived of thyroid hormone for a long time." He needed a different methodology. Instead of comparing normal maternal rats to thyroid-hormone-deprived rats, he began with a test group of pregnant rats, or "dams," that were hypothyroid to start with. He then timed his experiment for a very early stage in pregnancy, when the fetus would not yet have developed its own thyroid function, and thus fetal brain development might be dependent on maternal hormone crossing the placenta into the fetus. He compared a control group of hypothyroid dams with a hypothyroid group that had been given a single subcutaneous injection of thyroxine.

Using this new methodology, Amy Dowling, working under Zoeller's supervision, found eleven genes that were selectively affected by thyroid hormone, seven of which appeared to be enhanced by thyroxine and four that appeared to be suppressed by it. She went on to demonstrate that at least two of these genes coded for proteins that played an important part in brain development.

This was an important discovery, the first time in the scientific literature in which thyroid hormone was shown to regulate genes in the developing mammalian fetal brain.[5] It had clinical significance in suggesting that doctors wishing to avoid damage to the fetus needed to treat hypothyroid mothers, or potential mothers, as early as possible, before or during pregnancy. It also left an important question unanswered—a question that linked this work to evolutionary biology, and what had been learned about development from the study of metamorphosis. What, Zoeller now asked himself, was thyroid hormone really doing to the developing human brain?

The forebrain—what we commonly refer to as our cerebral hemispheres—is critical to human intelligence, morality, and sentience. The cerebral hemispheres are wrapped around big fluid-filled chambers, known as the lateral ventricles, and the developmental centers that construct the forebrain during embryology lie close to the lateral ventricles. The neurons that will populate the cerebral hemispheres arise in these developmental centers, and from here the developing cells migrate outward toward the convoluted brain surface we call the cerebral cortex. While still within the developmental centers these cells undergo eleven rounds of reproductive divisions, eventually populating the entire cerebral cortex in a series of six sequential layers. As the cells first proliferate, they express the beta-type thyroid-hormone receptor in their genomes. But as they leave the proliferative zone and migrate outward, they turn off the beta receptor and turn on the alpha receptor. To Zoeller, that switch has to mean something. The number of reproductive divisions is not affected by thyroid hormone. But when these cells begin to migrate, it seems that they are already committed to their eventual fate, whether to become a functional nerve cell, or neuron, or to become a connective tissue support cell, or glia. This determination decides the absolute number of nerve cells that will eventually populate the cerebral hemispheres and must have important implications for brain function in the future individual.

The key question, then, is what decides the cells' fate? Zoeller's researches suggest that the thyroid beta receptor, which is turned on at this time, appears to determine this. Once the cell's fate is decided, the beta receptor is switched off, and the alpha receptor is switched on to play a role that has yet to be determined.

If thyroid hormone, through interaction with its various receptors, is playing such an important role in human brain development, how, at the molecular level, is it doing so? For Zoeller this question brings us right back to the role of scientists such as Don Brown and their work on metamorphosis in the amphibian.

We have seen how metamorphosis in the sea squirt tadpole, bony fish, and amphibian is controlled by thyroid hormone. In the African clawed frog, thyroid hormone not only controls the metamorphosis from the tadpole to the adult but also, as Zoeller points out, metamorphosis takes the form of an orderly sequence of events, so that different tissues undergo thyroid-hormone-dependent metamorphic changes at different times and at different rates, in spite of the fact there is always a similar level of thyroid hormone in the blood. This suggests that there must be some system of control of the cellular levels of the hormone independently of the levels of hormone in the blood. It also suggested there might be some peculiar property of thyroid hormone when compared to most other hormones—a property that might explain why thyroid hormone has remained such a key control factor in development throughout the vast evolutionary time period of the chordate lineage.

In the 1990s Kathryn Becker and her colleagues demonstrated that local intracellular levels of active thyroid hormone, known as T3, did indeed vary even when the blood levels remained the same, and that such local variations were controlled by the actions of two opposing enzymes, one of which generated active T3, and the other which degraded active T3.[6] The counterposing actions of the two enzymes made possible a binary system of control of the thyroid hormone within the tissues. This simple but elegant system offered the means of controlling the timing of the action of thyroid hormone during development. In the frog *Xenopus laevis* metamorphosis from tadpole to frog involved major changes in eye positioning between tadpole and frog and in the visual fields perceived by the eyes. These were accompanied by the formation of a new nerve projection from the retina to the brain. In 1999, Donald Brown and his colleagues showed that these dramatic visual and neurological developments were indeed under the control of the two opposing enzymes.[7]

Just like frog metamorphosis, embryonic development of the mammalian brain takes place in an orderly sequence of events, and the timing of these maturational events is similar among different mammalian species.[8] In 2004, M. H. Kester and her colleagues reported that the same binary system of control of local thyroid hormone occurred in the cells of the fetal human brain during development, notably in the cerebral cortex.[9] As in amphibian metamorphosis, the amount of active hormone in the developing human fetal brain derived not from the levels of circulating blood levels of thyroid hormones, which were much lower than in the adult human, but from differential activity of the opposing enzyme systems within the cells. T3 levels increased in the developing cerebral cortex between weeks 13 to 20 until it reached levels higher than those found in the adult cortex. In contrast, T3 levels in the balance organ, or cerebellum, were very low at this time and increased only after midgestation, at 20 weeks. A similar enzyme-related system of local bioavailability of T3 was found to operate in the basal ganglia, the brain stem, the spinal cord, and another midbrain region known as the hippocampus.

"Would you say that thyroid hormone is essential not for the development of the brain per se," I asked Zoller, "but for the development of a *normal* brain?"

"Yes, that is true. A good example is one of the genes that Amy pulled out—it's a gene called HES, which stands for 'hairy enhancer of split.' This is a gene that essentially blocks neurogenesis. Now, it doesn't block proliferation, but it blocks the differentiation of an undifferentiated cell to a neural lineage. Instead these cells become glia."

"That's a terrific realization, isn't it?"

"Yes, it is. We've got a paper that came out recently on this. We're kind of making a proposition—it's a very difficult thing to study in vivo—but we're proposing that thyroid hormone controls the balance of the early production of nerve cells and glia. And that to me suggests that without thyroid hormones, you have the same number of cells as before: it's just that the balance between the two kinds of cell is lost."

To put it into simple words, what Zoeller, and these other researchers, have discovered is that this system, involving the opposing enzyme systems that control intracellular levels of active thyroid hormone—a system that evolved as a master control mechanism throughout the hundreds of millions of years of vertebrate metamorphosis—also controls the absolute numbers of nerve cells that go to form the cerebral hemispheres of the human fetus. It now seems inevitable that thyroid hormone played an important role in human evolution, especially the evolution and continuing development of the brain.

Truly we can now see that, in metamorphosis, what we are really witnessing is the wonder of development. From the fertilized egg to a complex human adult, we witness that same wonder. And now that we grasp the commonalities, our vision opens onto a window of more universal understanding of the wonder of life, in all the serendipity and linkages of its remarkable evolutionary story—its beauty inextricably one with its mystery.

— EPILOGUE —

A Sting in the Tail

Evening has overtaken me, and the sun has dipped below the horizon of the Ocean, yet I have not had time to tell you of all the things that have evolved into new forms.

—Ovid, *Metamorphosis*

Vincent B. Wigglesworth's contribution to our understanding of metamorphosis shines like a beacon through the pages of this book. This work was largely completed while he was working at the London School of Hygiene and Tropical Medicine, prior to his moving to Cambridge, to take up the Quick Professorship of Biology. But his wider contribution to insect physiology had only just begun. In a lifetime that began in 1899, when Queen Victoria was still on her throne, and which spanned almost the entirety of the twentieth century, sixty years of this spent in productive research, the great entomologist radically changed the way we think about the life of insects, their development—and more broadly—the physiology and development of life.

He was "the father of insect physiology," according to one of his students, Michael Locke, now professor and chairman of zoology at the University of Western Ontario, Canada, who, in 1995 Locke wrote an obituary on Wigglesworth for the Royal Society in London.[1] In correspondence with Locke, I asked him if he thought it reasonable to call Wigglesworth the "father of insect physiology."

"My authority for this," he replied, "would be Carroll Williams. He told me that when the volume [*The Principles of Insect Physiology*, first published in 1939] first came out, he was sitting with colleagues in a discussion group [at Harvard] formed with the express purpose of going through the book, and they were awed by what they read. It was what they had been waiting for—a book that systematized insect physiological knowledge to create a coherent science."[2] It was for the same reason that, in that same year, Wigglesworth was elected to fellowship of the Royal Society.

At Cambridge, where he spent the final two-thirds of his working life, Wigglesworth's genius was to lift the study of insect development beyond any applied scientific importance, admirable though this might be, to the level of a more universal understanding. To such a scientist, his ideals, the inner forces that drove and inspired him, would appear every bit as noble as the creation of the finest work of literature, or any other of the arts. John S. Edwards of the University of Washington's zoology department tells of how a single Wigglesworth paper inspired his career: "In a small colonial Zoology Department where dusty vertebrate comparative anatomy reigned, my introduction to the seeming simplicity and tractability of the insect, as revealed by Wigglesworth's paper, with its clear, direct prose and elegant spare line drawings, was the conversion experience."[3] He goes on to quote the American poet Robert Frost, who said that a poem should begin with delight and end in wisdom. "On that basis it might be claimed that many of Wigglesworth's best papers were poems of development."

For some, Wigglesworth's approach has come to represent a lost purity of ideal and endeavor. In April 1997, Michael Locke wrote despairingly of how modern science had entered a perilous state, threatened by "herd mentality" and "media-star" scientists. Locke felt so strongly about this that he joined forces with Peter Lawrence from the MRC Laboratory of Molecular Biology in Cambridge to publish a commentary to this effect in *Nature*.[4] Locke and Lawrence are not the only modern scientists to honor Wigglesworth for his dedication, humility, and his extraordinary contribution to learning. In 2002, some eight years after his death, Serap Akman of Yale University School of Medicine and his colleagues named a genus of bacterium that lives in the gut of the blood-sucking tsetse fly in his honor.[5]

It is unlikely that any such honor will be afforded Donald Williamson in his lifetime. Science is conservative by nature, and it is almost inevitable that the majority of scientists will mistrust iconoclasm. But iconoclasm is also important for the advancement of science: without it we would still believe the sun orbited the earth and that the continents have always occupied the positions they do today. Iconoclasm, in its turn, operates by challenging orthodoxy with novel hypotheses and theories. In the introduction to his seminal book on the structure of scientific revolutions, Thomas S. Kuhn explains how people have been misled by an overly simplistic attitude as to how scientific theories develop.[6] Most of us, whether scientists or nonscientists, come across theories through reading textbooks, or through lectures at colleges, or as established ideas in articles and communications, long after the theory was first conceived. In Kuhn's words, "the aim of such [communication] . . . is persuasive and pedagogic; a concept of science drawn from them is no more likely to fit the enterprise that produced them than an image of a national culture drawn from a tourist brochure."[7] Williamson's story, whether one agrees with his theory or not, is illuminative of the concept of ideational evolution.

We have glimpsed, at various stages during its development, how, following first conception, Williamson's initial hypothesis of larval transfer began to change and expand, evolving, through the confirmation of experiment, from hypothesis to theory. "Progress in knowledge," as Stephen Jay Gould once explained, "is not a tower to heaven built of bricks from the bottom up, but a product of impasse and breakthrough, yielding a bizarre and circuitous structure that ultimately rises nonetheless."[8]

In mid-January 2002, as Williamson led me on an inspection tour of the tank-holding facilities at the marine laboratory, we were discussing the evolution of his theory. We walked across wooden ramps between a variety of holding facilities in the old and somewhat dilapidated indoor labs—already heralding a still gloomier fate—where I marveled at spider crabs looking back at me from their perches on stones, and scallops within their shells, like perfect representatations of the Shell Oil sign, starfish, sea urchins, octopuses—in my ears the sounds of rushing water, while my nostrils were overwhelmed with the briny smell of the oceans.

By that time, I had come to agree with Williamson that there seemed to be many anomalies in marine metamorphosis, anomalies that were difficult to explain along orthodox evolutionary lines. I knew this perception had led him to construct a novel if highly controversial explanation, and one that was still continuing to evolve even then. His theory, insofar as I was aware, was the only one that attempted a universal explanation of metamorphosis throughout all of the animal kingdom. This in itself seemed worthy of exploration, and so I asked him the inevitable question:

"You wrote the theory in a series of papers and then in book form, in 1992. But even after that, you subsequently expanded the theory in a really big way. I just wondered why you felt you needed to do that."

"Well, there was no dramatic breakthrough. It was very gradual. I realized when I wrote the original paper, and the book, that this was not the end. There must be lots more cases that I had not yet considered. I started to consider the crustaceans in greater detail. I was already a crustacean specialist. I decided that most crustacean larvae had been transferred from other groups. I was still, as I then saw it, hanging on to the probability that creatures that metamorphosed in a more gradual way were part of the same genome as the adult, whereas those that showed a drastic metamorphosis certainly were not. A crab hatches out as a so-called zoea larva, which is nothing whatever like a crab. On the other hand a megalopa larva is quite like a crab.[9] [The megalopa is often, but not always, the final larval stage]. Yet the metamorphosis of zoea to juvenile crab is very gradual. At that stage I thought that the megalopa was part of the original genome, and the zoea was from another source entirely. I would now say that the megalopa is also from another source, another vaguely crablike animal that doesn't need a big metamorphosis. I then wrote a paper on the evolution of crustacean larvae. I could not draw for myself so I got Tony Rice to provide the illustrations. I sent it in to *Crustaceana*, the specialized journal, and I was very surprised when they accepted it, almost immediately and without a quibble.[10] Anyway, from that time on I was extending it gradually with the idea that nearly all larvae had been acquired from other sources, but I was hanging on to the possibility that it wasn't quite all."

In the laboratory I found myself gazing at "Dead man's fingers," *Alcyonium digitatum*, shaped like a stumpy white and bloodless hand with numerous sausage-shaped fingers. These are coral-like animals that, when feeding, produce numerous tiny polyps, giving their surface a fuzzy appearance thought to resemble decaying flesh. Nearby, I almost crunched my feet on some "mermaids' purses," quadrangular shapes about two inches long, brown and gelatinous, the eggs of both dogfish and many species of sharks.

Our conversation continued as we left the indoor laboratory and walked round the back of the building to where some large outdoor tanks were set down in the shadows of rocky cliffs. Gulls wheeled and cried above our heads. Williamson pointed out a bricked tank built into the cliff. This had been his source of *Ascidia mentula* up to 1990, but it was leaking even then, and soon afterward it was judged beyond repair. I found myself gazing up at the huge concrete-walled tank that had replaced it, currently being used to spawn fish. From here the larval forms or small fish were being delivered to people who needed them to restock marine areas. This same tank would soon provide the *Ciona intestinalis* and *Ascidiella aspersa* for the 2002 experiments; for further studies of *Ascidia mentula*, divers would have to journey out and explore a shipwreck three miles northwest of Bradda Head.

In a series of papers, written over the previous decade or so, Williamson had extended his theory. At the time of our conversation he was working toward the publication of a second book, *The*

Origins of Larvae,[11] in which he would extrapolate his theory of larval transfer to embrace all forms of marine invertebrate larvae and, in subsequent years, in what would prove to be the most controversial expansion, he would further extend it to include metamorphosis in insects.[12] In his early work, and first book, however controversial it might have seemed to colleagues, he had confined himself to his chosen discipline of marine biology. He had put forward a consistent theory—for example, the radical shifts in axes of development and symmetry, from bilaterian to radial—and he had devised and conducted experiments aimed at confronting that theory. This latter extension seemed to me less consistently logical, and in including insects it took him away from his own scientific discipline. Unlike his earlier theory, he had performed no probing experiments involving insects. I questioned this latter extension: "I'm still not quite clear in my mind why you thought that was the case."

"Well, I was reluctant at that time to go the whole way. When I wrote the original paper I was more or less convinced that [larval transfer] applied only to the exceptional cases. Other larvae had evolved from the same source. The crustacean paper was a big breakthrough. This was followed by the Fallen Leaf Lake Conference."

In September 1996, roughly four years after publication of his first book, Williamson had been contacted by Michael Syvanen, a professor of medical microbiology and immunology at the School of Medicine, University of California–Irvine, to contribute to a conference on horizontal gene transfer at Fallen Leaf, California. He described the context for me:

"I don't know how I got into contact with Mike Syvanen. He's the horizontal gene transfer man. Lots of work had been done on bacteria, and there was no doubt that genes could be transferred between bacteria; he was delighted to get an example of a theory of gene transfer in eukaryotic animals. So he invited me to his conference in California.

"Enid and I hemmed and hawed about whether we could get there. Eventually we decided that we couldn't. So I went to Liverpool, to the university, and I got them to make a video of the theory as it then was. The video did not go much further than my 1992 book. I sent them a copy of this, which went down very well. . . . Then I was invited to write an article on the same theory for the book that followed. I started off more or less writing what was in the video. In the meantime I had written the crustacean paper. This was already in press. I was now happy with the idea that larval transfer was much wider in application than in the original paper or the book. Each time I sent Syvanen my manuscript, I would follow this with a letter to say that a revised version would follow. Each version traced another larval form to an adult in another taxon: trochophores to rotifers resembling *Trochosphaera*, and tornarias to an ancestor of *Planctosphera*. This meant that Johannes Müller's larva is not the primal trochophore, and acorn worms did not invent their own larvae. All larvae originated as adults in foreign taxa. The horizontal gene transfer book was published in 1998, with a revised edition in 2002."[13]

"By that time I had decided that all drastic metamorphoses were due to the larval form being acquired from another group. But I was still undecided about the more gradual metamorphoses. It was while I was writing the chapter in Syvanen's book that I got the idea that this could be extended further, not only in the crustaceans but also to different groups."

"You kept extending the application of your theory?"

"Yes. I kept extending it. I eventually reached the stage where I felt I could extend it not only to all larvae but to all embryos. Again, it just seemed to fit, so why not? There is no other explanation for where embryos come from, so why not larval transfer or some similar horizontal genetic transfer."

We have seen that at the heart of biological theory there is a great divide on the origins of metamorphosis. On the one hand we have the Haeckelian theory, in modern form best exemplified by the Swedish biologists Gösta Jägersten and Claus Nielsen, which takes the view that metamorphosis is ancestral to the evolution of all animal life history. On the other hand, we have the direct development theory, proposed by Williamson, in which the earliest animals did not metamorphose. In the former the larval to adult cycles evolved at the very beginnings of animal life and led, within the same individual life cycle, to a free-swimming "pelagic" phase (the waters above the ocean bottom) that would become the primary larva, and to a bottom-dwelling "benthic" phase that would become the adult. This theory suggests that there were survival advantages in a dual life cycle, with the pelagic forms better equipped to disperse to wider distances and new habitats in the three-dimensional ecosphere of moving oceans, while the benthic forms were better equipped to explore the rich food sources and other potentials of the two-dimensional ecosphere of the ocean floor. The Haeckelian theory makes two important assumptions. It takes for granted that "primary" larvae evolved through a process of linear descent with modification from the earliest and simplest of animal forms, dating back to the Precambrian age. It also implies that all the complex life cycles that we see today must have come about as a result of linear descent with modification of that primal cycle.

Is this a reasonable working theory?

In my opinion this theory will founder if it assumes that the only mechanism of evolutionary change is exclusively that of mutation under the direction of natural selection, which has very often been the assumption of the past. But if we enlarge this theory to embrace all of the mechanisms for hereditary change, including mutation, hybridization, genetic symbiosis, and epigenetics—and if we include the operation of such mechanisms during embryonic development—with natural selection working at different levels, then it becomes a good deal more reasonable. Would this theory unite the world of evolutionary biology? Based on the analysis of Richard Strathmann, and the continuing debate at large in the world of developmental biology, it probably would not. It would struggle to explain all instances of metamorphosis—for example, an original metamorphic life cycle could not explain the origins of complete metamorphosis in insects, since most entomologists are agreed that metamorphosis was not ancestral to insects. It evolved long after the ancestral forms of the first insects invaded dry land. Moreover, a significant number of marine biologists also believe that many of the larval forms seen today in marine invertebrates also evolved after the adult forms had already evolved.

If Truman and Riddiford are right, the breakthrough to the dramatic change of complete metamorphosis in insects may have come about when one group evolved the ability to hatch out early, releasing an offspring at a premature but nevertheless viable stage in its embryonic development. What, in such a scenario, might have given rise to an inherited pattern of viable premature hatching? One possibility is a mutation, or possibly a series of mutations, affecting developmental pathways. But it is important that we continue to consider the other mechanisms itemized above: not merely mutation, but also hybridogenesis, symbiogenesis, and epigenetics, even if we accept the basic Truman and Riddiford hypothesis.

Williamson refutes all such extrapolations of the Haeckelian theory. He disagrees with the theory that larvae originated in a linear descent-with-modification fashion as part of a primal pelagic-waters evolutionary phase and an ocean-bottom benthic phase, all within the same life cycle. Instead he proposes that some animals evolved an entirely pelagic life cycle while others evolved entirely benthic cycles. In his view, marine metamorphosis came about not within a single animal's life cycle but through hybridization between two such dissimilar life cycles, resulting in a sudden evolutionary leap—in evolutionary jargon, a saltation—that combined the pelagic and benthic life cycles of the dissimilar parent species within a single descendent organism.

It is important to realize that, in his view, this did not happen in the recent past. On the contrary, he proposes that this happened long ago, when genomes might have been less honed by selection. The frequency and diversity of hybridizations would undoubtedly have been facilitated by the broadcast spawning so prevalent in marine life-forms. How varied must have been the outcomes in terms of new animal forms and lifetimes through millions of putative hybridizations over such vast time periods with long-term success or failure rewarding those with the greatest selective advantage not only in terms of form but also in terms of life cycle. In his experiments at cross-phyletic hybridization, he realized that it was perhaps too much to hope that over a small number of experiments he could recapture an occasional event from the very distant past. All he really dared to hope for was that that some form of hybridization would occur, and be recordable.

We need to ask ourselves not merely if Williamson's theory of metamorphosis by hybridization is true, but whether his theory, and work, contributes, specifically or generally, to scientific understanding.

We might recall Sebastian Holmes's response when I asked him what was needed to confer reasonability on Don Williamson's theory. Holmes felt Williamson had already done enough to confer reasonability, but, together with Nic Boerboom, Holmes had combined forces with Williamson to confirm those earlier experiments by performing the new hybrid experiment, a cross-phyletic sexual union that resulted in a previously unknown asexually reproducing life-form. This experimental result has not, to date, been confirmed by genetic analysis of the hybrid offspring. Further study is surely warranted. Moreover, none would disagree with Richard Strathmann that such further study must include rigorous molecular and genetic analysis—not a superficial examination of one or two genes, but a rigorous and meticulous analysis, stage by stage, at the level of painstaking detail pioneered by Rudolph A. Raff.

Such genetic and molecular studies should readily confirm or refute that the spheroids are genuine cross-phyletic hybrids. And if confirmed, much might be learned of unknown genetic systems of genetic and epigenetic control. Indeed we simply do not know what to expect—and that in itself might make the cost and effort of such a major undertaking eminently worthwhile.

There is an additional pressing reason why we should consider Williamson's broader viewpoint. In the words of the late Stephen Jay Gould, "The pleasure of discovery in science derives not only from the satisfaction of new explanations, but also, if not more so, in fresh (and often more difficult) puzzles that the novel solutions generate."[14] And it is in consideration of Gould's interesting premise that I shall close this account of Williamson's strange, exciting, and always probing exploration, this time with his extension of hybridization and its role in metamorphosis to help explain a second mystery, one less familiar to lay readers than that of metamorphosis, yet one that has loomed over the entire world of zoology for a century or more. This is the mystery of the origins of the animal kingdom, known to biologists as the Cambrian explosion.

The Burgess Shale is a fossil bed dating back to the Middle Cambrian age, some 545 to 525 million years ago. In this fossil bed, and an increasing number of related deposits elsewhere throughout the world, the forerunners of modern animals appear for the first time. As has become increasingly clear, their arrival in the fossil record was not heralded by a vast period of progressive evolution of the various animal groups. Instead there appears to have been an amazingly rapid and extraordinary flowering of different branches of the tree of life, during which, as Gould expresses it in his book, *Wonderful Life*,[15] "modern multicellular animals make their first uncontested appearance in the fossil record . . . with a bang, not a protracted crescendo." This bang is known as the Cambrian explosion.

For the first three and a half billion years that life existed on Earth, it remained at the very basic form of bacteria. By 2.5 billion years ago, bacteria had evolved the ability to capture the energy of sunlight through photosynthesis, ushering in the Proterozoic era, which would extend to the appearance of the first multicellular animal forms during the Vendian period, thought to have extended from 600 million to 545 million years ago.[16] Vendian forms originally related to Precambrian fossils found in the White Sea area of Russia, just north of Archangel, while another group of early animal forms, known as the Ediacaran fauna, originally referred to slightly earlier Precambrian fossils found in the Ediacaran Hills of Australia. Today there is a confusing tendency to mix terms and dates, but even if we broadly attribute the Precambrian as starting about 600 million years ago and the Cambrian about 545 million years ago, the intervening period, perhaps some 55 million years, is, in terms of geological time, a very short interval for virtually all of the animal phyla to make their first appearance.

Some of the earliest Precambrian fossils are little more than blobs that are difficult to attribute. Others appear to represent early cnidarians, like the present-day jellyfish, sea pens, and sea anemones. Among the many enigmatic fossils discovered in Ediacaran rocks is an impression of an animal named *Arkarua adami*, named for Arkarua, the mythical giant snake of the Aboriginal peoples who live close to where the fossil was discovered, in the Flinders Ranges of south Australia.[17] *Arkarua* is a small disklike fossil with a body that is divided into five lobes—pentaradial. It may represent the earliest echinoderm, which, if true, would confirm Williamson's contention that the first echinoderms were radial.[18] Unfortunately, it lacks soft-tissue details, so the link to the echinoderms can only be speculative at present.

In a much broader respect, the Cambrian explosion gives added credibility to Williamson's thinking of the importance of hybridization both in marine metamorphosis and animal evolution. Suddenly, from a geological perspective—and Simon Conway Morris, professor of evolutionary paleobiology at the University of Cambridge, brackets the principal events of evolutionary interest as taking place between 550 and 530 million years ago—a massive diversity of animal forms appear in the fossil record.[19] In geological terms, this is truly dramatic. Others, who have reported on the fossil record of a contemporaneous Chinese site, known as the Chengjiang biota, have looked specifically at the development of animals with exoskeletons and the chordate beginnings of endoskeletons, and have proposed that most of these forms appear over a period of just ten million years.[20]

Any universal theory of evolution is obliged to explain how, over such a brief period of time, the body forms of most of the present-day, and many extinct, animal phyla come into being. But no satisfactory explanation has ever been made. On the contrary, the Cambrian explosion has been a source of controversy since first observed by geologists in the 1830s. Charles Darwin was well aware of its challenge to his ideas of gradualist evolution, and he devoted the ninth chapter of *The Origin of Species* to what he termed "the imperfection of the geological record," concluding that if his theory were true, it demanded that a period of gradual evolution must have preceded the Cambrian explosion, "as long as, or probably far longer than, the whole interval from the Silurian [Cambrian] age to the present day; and that during these vast, yet quite unknown periods of time, the world swarmed with living creatures."[21] He added, "To the question why we do not find records of these vast primordial periods, I can give no satisfactory answer." Today there is no escaping the fact that Darwin was mistaken. The apparent explosion is not the result of imperfections of the fossil record. The much better fossil record we have today strongly suggests that the Cambrian explosion is real.

The animals fossilized in the Burgess Shale do not lie within the Cambrian explosion itself, but they appeared soon after it, in Gould's words, "before the relentless motor of extinction had done much work." This lends them, and similar groups discovered in contemporaneous sites worldwide, enormous importance. Moreover, their preservation, like that of the Chengjiang biota, is all the more precious in revealing not only fossilized skeletons but also the most exquisite detail of soft-body forms. While many of these forms are ancestral to modern animal phyla, a number of forms bizarre and improbable to our conditioned vision are also preserved, with no readily discernable legacy in the living animal phyla. These Vendian and Cambrian fossils offer an extraordinary snapshot of animal evolution during and very soon after its early diversification. Perhaps we should consider what we are privileged to witness here: not circus monsters with names such as *Hallucigenia* and *Anomalocaris*, but the true wonder of life at its most creative—forms that suggest intriguing evolutionary experiments, arising from what Gould describes as a "staggeringly improbable series of events . . . utterly unpredictable and quite unrepeatable."[22]

One such creature I mention in passing. *Pikaia gracilens* is a slender lanceolate creature about an inch and a half in length, with a fishlike musculature, the presence of a notochord, and a peculiar-shaped head, which divides into two fleshy extensions, or lobes. Conway Morris believes that this animal, like the 15-million-year-older *Cathamyrus diadexus* from the Chengjiang deposits, must be one of the earliest chordates.[23] *Pikaia* may have been a filter feeder that swam through the water with undulations of its body, much like a fish. Perhaps a more general lesson in early animal evolution might be gleaned from examining some of the more "improbable, unpredictable and unrepeatable" creatures.

Marrella splendens is the commonest fossil animal found in the Burgess Shale, with more than 15,000 examples catalogued to date. A tiny animal no bigger than a baby shrimp, it has a striking appearance with a shield head decorated with two broad horseshoe-shaped antennae. Its body has between 24 and 26 segments, each sprouting pairs of branched appendages. The upper branches are long feathery gills for breathing and the lower branches comprise 24 to 26 pairs of legs. Its mouthparts are unknown. When Charles Walcott, an eminent American paleontologist, first discovered it, he thought it represented an extinct species of trilobite, one of the commonest animals seen in the fossil record but which today are extinct. But Harry B. Whittington, a professor of geology at the University of Cambridge, reappraised it in 1971 and concluded that it just did not fit into any of the known modern phyla.[24] Today *Marrella* is seen as a unique category within the superphylum of the arthropods.

Another strikingly beautiful animal, *Burgessia bella*, is named for the fossil site itself. *Burgessia* is the third commonest arthropod found in the Burgess Shale. Seen from above, it is encased in an oval carapace, somewhat resembling the head shield of a horseshoe crab, though only a fraction of its size, with two delicate curled antennae projecting from the front end and from the rear an elongated spike of a tail, twice the length of the carapace. Underneath the carapace, the body is divided into two distinct parts. Up front, and immediately behind the sprouting antennae, is a mouth devoid of jaws. To date, paleontologists have found no definite evidence of eyes. The front part also bears three sets of legs. The rear part bears seven branched appendages, each dividing into outer gills and ground-facing legs. In the words of C. P. Hughes, who reexamined it in 1975, "What is apparent from this restudy is that *Burgessia* possess a mixture of characters . . . many of which are to be found in modern arthropods of various groups.[25] Gould remarks that Hughes felt unable to place *Burgessia* in the standard taxonomic tree, "because he regarded this genus as a peculiar grab-bag, combining features generally regarded as belonging to a number of separate taxonomic groups."[26]

In his charming book *The Crucible of Creation*,[27] Simon Conway Morris describes another creature, *Nectocaris pteryx*, with exquisite detail since he had dissected its single known fossil from the Burgess rock.[28] What Simon Conway Morris sees in the fossil is a very peculiar creature indeed, with enormous eyes right in front of a carapace-shrouded head. But if the head is reminiscent of that of an arthropod, the abdomen appears to belong to an entirely different phylum, with prominent longitudinal fins and fin rays. In his words, "Such a feature is never seen in the arthropods."[29] In the modern oceans, arrow worms are speedy and efficient hunters, with streamlined bodies similar to that of *Nectocaris* and fins supported by fin rays. But *Nectocaris* bears little resemblance to an arrow worm. The simple truth is that its front end resembles an arthropod while the remainder resembles a chordate, complete with tail and fin. In Gould's words, "What can be done with such a chimera!"[30]

Again and again, in the reappraisals of animal fossils from the Cambrian explosion, we encounter what Williamson would describe as significant anomalies. In Gould's words: "Thus, the bivalved arthropods—the group that seemed most promising as a coherent set of evolutionary cousins—also formed an artificial category hiding an unanticipated anatomical disparity." What possible order could be found in the Burgess arthropods? "Each one seemed to be built from a grab-bag of characters—as though the Burgess architect owned a sack of all possible arthropod structures, and reached in at random to pick one variation . . . whenever he wanted to build a new creature."[31]

In the opinion of James W. Valentine, professor emeritus at UC Berkeley and a distinguished evolutionary paleontologist, the Cambrian explosion must reflect "some very special circumstances in life's history."[32] It is hardly surprising that it has provoked a remarkable range of speculative explanations—indeed so remarkable and varied that when Valentine tried to organize all such hypotheses into a comprehensive framework, he was unable to do so. Instead he rounded it all off as best he could to a small number of representative ideas: some biologists merely dismissed the reality of the Cambrian explosion; others attributed it to physical changes in the environment, including rapid rise of atmospheric oxygen; others built hypotheses based on biological changes in the environment, such as the consequences of the rise of new planktic forms and the evolution of predators; and others still put it down to what Valentine labeled "intrinsic evolutionary change." Surely common sense would suggest that, whatever the contribution of the physical environment or the biological component of the ecology, any reasonable explanation of the Cambrian explosion must come from the known mechanisms of evolutionary biology.

In his second book, *The Origins of Larvae*, Williamson highlights another anomaly, this time not larval but appearing in the body plan of several groups of animals known as "lophophorates." All lophophorates have a circular or horseshoe-shaped ring of tentacles they use to trap their food. The lophophores include four marine invertebrate phyla: the bryozoan, or moss animals, which are tiny filter feeders; the phoronids, which are bottom-feeding worms; the brachiopods, also known as lampshells because they resemble ancient oil lamps in having their bodies hidden between two unequally sized clam-shaped shells; and finally, the entoprocts, tiny transparent marine animals previously classed with the bryozoans but excluded these days because the anus as well as the mouth opens within the lophophore.

Within the phylum of the hemichordates—we might recall that they are a group of headless chordates that includes the acorn worms—the pterobranchs also feed using lophophores. In Williamson's opinion, the lophophore is such a complex organ it is unlikely to have evolved more than once. And this, in turn, would suggest that the various, seemingly unrelated groups of animals that possess a lophophore—the lophophorates—must in fact be related to each other through some common evolutionary ancestor. Pat Willmer, a distinguished biologist at the University of St. Andrews, acknowledges that the lophophorates seem to belong together and to stand somewhat apart from the other members of the phyla: "Most importantly, they show a very complex mixture of the set of features that are normally used to define protostomes and deuterostomes, thus giving them an apparently anomalous position somewhere in the middle of the animal kingdom." She even conjectures that the lophophorates might be "the transitional stage in the 'invention' of the deuterostomes."[33]

In ruminating over this quandary, Williamson wondered if these difficulties could be reconciled if the lophophorate developmental plan had been transferred between taxa in the distant past through hybridization?

If so, it would open the floodgates to many other possibilities. What if there had been sporadic hybridizations between organisms ever since the invention of sex—an event that long preceded multicellular animals? Sexual crossing between disparate species of animals at a very early stage in their evolution might have resulted in the blending of body forms and components. The resulting offspring, and their descendants, would become chimeras—amalgamations of quite different species. In his words, "the more I thought of it, the more it made sense." This new line of thinking inevitably brought him face-to-face with the enigma of the Cambrian explosion.

In April 2004, at the invitation of Lynn Margulis and Wolfgang Krumbein, he spoke about his new ideas at a biological conference in Bellagio, Italy.[34] Here he talked about the expansion of his hybridization theory to lophophorates for the first time. In December of the same year, Margulis suggested that Williamson should contribute to an online cultural event, "The Edge New Year," aimed at broadcasting cusp-of-the-wave debate and enlightenment, and it set out to challenge each New Year with a broad but seminal question to which various thinkers might freely contribute a personal answer. The Edge New Year question for the opening of 2005 was "What I believe but cannot prove."[35] In Williamson's own words, "I said, in half a page, that I believed I could explain the Cambrian explosion."

What prompted the marine biologist to make such an announcement? Valentine, in appraising the first appearance of what we now recognize as phyla in the early Cambrian fossil record, was convinced that the fossil record really was consistent with the actual evolutionary events. In his book *On the Origin of Phyla* he states: "The abrupt appearance of disparate body types without a record of intermediate forms . . . suggests that significant morphological change may evolve very rapidly."[36] In case his readers should harbor any lingering doubts, he emphasizes what this means: "The implication is that the abrupt appearance of many phyla during the [early Cambrian] could mirror an actual rapid radiation of body plans, just before or during that time period." This view is widely shared among evolutionary biologists. The question then is how such major body changes could evolve in such a brief time period? While we can reasonably assume that all of the mechanisms of genomic creativity will have played some part, one such mechanism, hybridization, is an important source of rapid, large-scale change. And now this force, so long neglected as a source of change in animal evolution, sprang to Williamson's mind.

Williamson, like most biologists, had long been puzzled by the mystery of the Cambrian explosion. "Not until the summer of 2004 did I read Stephen Jay Gould's *Wonderful Life* on the Burgess Shale fauna. Now I decided that these remarkable animals were probably the results of component transfer—they were concurrent chimeras." Williamson wrote a new paper, with the challenging title, "Hybridization in the Origin of Animal Form and Life-Cycles."[37] In this paper, published in 2006, he extended his hybridization theory to embrace both larval transfer and the Cambrian explosion, developing his ideas along lines that will now be familiar. The early animals were exclusively marine. They shed their eggs and sperm into the water, where fertilization took place. At such a time, when genomes were presumably more malleable to hybrid union than today, hybrid unions must have been common, not only between the first animals but also between the resulting hybrids. "This produced many concurrent chimeras, which included the first members of most animal phyla, living and extinct."

Williamson acknowledges the importance of additional factors that might have encouraged genomic experimentation at the time of the Cambrian explosion. Paleontologists believe that the Cambrian explosion followed the great Precambrian extinction, dated at about 650 million years ago, and possibly the greatest devastation of life on Earth. The same event may have led to the catastrophe known popularly as the "Snowball Earth," when approximately 70 percent of the Precambrian flora and fauna appear to have perished. Some biologists believe that this was followed by a lesser extinction, at the end of the Vendian, which may have wiped out a great many of the survivors from the earlier extinction. Whatever the cause, or causes, of these catastrophes, a great many ecological opportunities may have arrived in their wakes, presenting themselves to the newly evolving animals of the early Cambrian. Other authorities, such as Valentine and Conway Morris, have suggested the contribution of exceptional availability of food and nutritional sources to a new proliferation of life.

There is some evidence that the oceans, in early Cambrian times, contained relatively high levels of phosphorus, a compound that modern-day oceanic and terrestrial farmers would be altogether familiar with as a prerequisite for rapid and healthy biological growth.[38] Natural selection would have determined the fate of the many contemporaneous genomic experiments, leading to the animal phyla we are familiar with today. Others that subsequently became extinct survived long enough to be captured, in spectacular and beautiful detail, in the Cambrian fossil record.

In 2010, proposing that that the same force, hybridization, was capable of explaining both the origins of larvae and the origins of the animal phyla during the Cambrian explosion, Williamson brought together the two interrelated strands of his theory in a new publication.[39] This extension of his hybridization theory to the origins of the animal kingdom soon proved every bit as controversial as his larval transfer theory. And exactly the same questions apply, in biological and evolutionary terms. Does the theory have sufficient validity to make further testing reasonable and worthwhile?

With the availability of whole genome sequences, Michael Syvanen, the horizontal gene-transfer expert, has begun a systematic evaluation of the evolutionary origins of the major phyla, including both plants and animals, reappraising the various basic groups and their places on the tree of life. In a paper published as long ago as 1985, when Syvanen was working at the Harvard Medical School, he proposed that horizontal transfer of genes between different groups, or taxa, across the evolutionary tree—already observed to take place between different groups of bacteria—was likely to apply to all of life. Such transfers would bring pre-evolved genes from different evolutionary lineages together in sudden evolutionary bursts, and this might help to explain, at least in part, the discovery of similar genes throughout all of life, as well as rapid bursts of evolution, such as the Cambrian explosion. In 2010, Syvanen and Jonathan Ducore, now working in the Departments of Microbiology and Pediatrics at the University of California at Davis School of Medicine, compared and contrasted the whole genomic sequences from some of the major animal phyla. Their conclusions challenged some of the prevailing assumptions.[40]

In essence, evolutionary theory predicted that phyla with closer evolutionary links to one another would inevitably have more genes in common. These similarities would also show if they examined proteins instead of genes, since protein sequences derive directly from genetic sequences. Based on the expanding knowledge of whole genomes, Syvanen and Ducore possessed the sequences of two thousand different proteins, and now, counting the numbers of proteins in each separate phylum, they could group the animal phyla into evolutionarily linked clusters—or "clades"—based on the numbers of shared proteins. Extrapolating from the conventional tree of life, such an exercise should confirm the evolutionary commonalities of the deuterostomes, including the sea squirts (also known as tunicates or ascidians), with their tadpole larvae, the sea urchins, with their pluteus larvae, and humans, as examples of the vertebrates. But when they looked for such evolutionary commonalities of the two thousand proteins they found that they fell into two distinct groups. About half the protein sequences supported the notion that the tunicates were indeed related to the chordates. However, the other half suggested that the tunicates descended from an ancestor whose closest living relatives are found amoung the arthropods and nematodes or with the protostomes—in other words a group that had no common ancestry with the chordates.

These results suggested that a major horizontal gene-transfer event had taken place during the emergence of one of the animal phyla. The simplest explanation, in the opinions of these authors, was that the ascidians began as a hybrid between a primitive chordate and some other organism, perhaps "from an extinct and unidentified protostome phylum, at a time close to but after the diversification of the chordates and echinoderms." As they further explained: "We are not the first to be intrigued with tunicate taxonomy. The tunicate anomaly has perplexed students of biology for more than a century and led Don Williamson to suggest that the tunicate had hybrid origins with those genes controlling larval development coming from a chordate ancestor, and those genes controlling adult development coming from some other phylum."

This research by Syvanen and Ducore is pioneering in its thinking and methodology, and their findings will be seen as controversial. Other scientists are likely to take up this innovative methodology to confront their findings—as indeed they should. But if, in time, Syvanen's and Ducore's conclusions are confirmed, it will demonstrate the importance of hybridogenesis as a mechanisms for major, and relatively rapid, evolutionary change at the very root of animal evolution. It will also illustrate the importance of iconoclastic thinking in science, the sort of thinking that has inspired Williamson's approach. This, however, does not mean that evolutionary science has been mistaken for years.

Mutation, in its original sense of genetic errors arising in the copying of DNA during cell division, remains important to understanding the genetics of hereditary change, and it continues to offer important understanding of the genetic underpinning of many diseases. But mutation is insufficient on its own to explain the full range of hereditary change, whether in evolutionary biology or in the genetic underpinning of disease. Hybridologists don't regard hybridogenesis as mutation, any more than symbiologists regard genetic symbiosis, or epigeneticists regard epigenetic change. Modern evolutionary biology, including the growing science of evo-devo, must surely embrace all of these mechanisms for hereditary change, as, through the work of a new generation of scientists, aided by increasing understanding of the complete structures of genomes, evolutionary science is coming to grips with some of the great mysteries of biology, including metamorphosis and the Cambrian explosion.

REFERENCES

Prologue: The Beautiful Mystery

1. Conard, N. J. (2004). Palaeolithic ivory sculptures from southwestern Germany and the origins of figurative art. *Nature* 426: 830–32. See also A. Sinclair (2004). Archaeology: Art of the ancients, *Nature* 426: 774–75; R. Dalton (2003). Lion Man takes pride of place as oldest statue, *Nature* 425: 7.
2. Hammond, N. Stone Age artists are getting older, Timesonline, February 13, 2006; J. Roach, Ancient figurines found—from first modern humans? *National Geographic News*, December 17, 2003.
3. Wigglesworth (1976): 168.

Chapter 1: The Birth of an Idea

1. All direct quotes from Williamson are from my interviews and various communications.
2. See online article by Russell D. Fernald at: http://www.karger.com/gazette/64/fernald/art_1_0.htm.

Chapter 2: A Puzzle Wrapped in an Enigma

1. Tattersall, W. M., and E. M. Sheppard (1934), Observations on the asteroid genus: 35–61.
2. Haeckel, E. (1874). Die Gastraea-Theorie, die phylogenetische Classification des Thierreichs und die Homologie der Keimblätter. *Jena Z Naturw* 8: 1–55.
3. Balfour (1880–81).
4. Darwin, F. ed. (1905). *The life and letters of Charles Darwin.* New York: D. Appleton & Co: 426.
5. Ibid.
6. Hall, B. K. (2000). Balfour, Garstang and de Beer: The first century of evolutionary embryology. *Amer Zool* 40: 718–28. See also, Hall, B. K. and M. H. Wake (1999).

Chapter 3: First Experiments

1. Darwin, C. (1871).
2. Barrington, E. J. W. (1968). Metamorphosis in lower chordates. Chapter 6 in W. Etkin and L. I. Gilbert, eds. (1968).
3. Meinertzhagen, I. A., and Y. Okamura (2001). The larval ascidian nervous system: The chordate brain from its small beginnings. *Trends in Neurosciences* 24 (7): 401–10.

Chapter 4: The Price of Iconoclasm

1. Fell, H. B. (1941). The direct development of a New Zealand ophiuroid. *Quart J Microscopical Science* 82: 377–441.
2. In the following paper, Kirk uses the old term "sand-star" for what is now known as a brittle star.
3. Kirk, H. B. (1916). Much abbreviated development of a sand-star (*Ophionereis schayeri?*). *Transactions of the New Zealand Institute* 48: 12–18.
4. Hyman, L. H. (1940–49).

5. Russo, A. (1891). Embriologia dell'amphiura squamata. *Atti R. Acad. Napoli*, ser. 2: 5.
6. Williamson (2003): 117.
7. Fell, H. B. (1968). Echinoderm ontogeny. In R. C. Moore, ed. (1968): S60–S85.
8. Rowe, F. E. W., A. N. Baker, and H. E. S. Clark (1988). The morphology, development and taxonomic status of *Xyloplax* Baker, Rowe & Clark (1986) (Echinodermata concentricylcloidea), with description of a new species. *Proc Roy Soc London*, B; 233: 431–59.
9. Schatt, P., and J.-P. Féral (1996). Completely direct development of *Abatus cordatus*, a brooding schizasterid (Echinodermata: Echinoidea) from Kerguelen, with description of perigastrulation, a hypothetical new mode of gastrulation. *Biological Bulletin* 190: 24–44.
10. Williamson (1992): preface.

Chapter 5: Challenging the Tree of Life

1. Williamson, D. (1987). Incongruous larvae and the origin of some invertebrate life-histories. *Prog Oceanog* 19: 87–116.
2. Darwin (1859), 1st ed.: chapter 13.
3. Ibid., 6th ed.: chapter 14.
4. Ibid., 6th ed.: chapter 14.
5. Willmer (1990): 123; see also p. 313.

Chapter 6: "This Is Impossible!"

1. Margulis (1970).
2. Williamson (1992): foreword.
3. Darwin (1859), 6th ed.: chapter 14.
4. Williamson interviews. See also Williamson (1992).
5. Williamson (1992): editor's comments.

Chapter 8: The Evening of the Great Peacock

1. Fabre (1912).
2. Darwin (1871).
3. Alcock (1997): 38.
4. Darwin letter to Fabre. http://www.e-fabre.net/virtual_library/correspondence/romanes.htm.
5. Legros, G. V. (2002). *Fabre, Poet of Science*. Reprint, Indypublish.com. This book, an English translation from the original French, is freely available online.

Chapter 9: The Science of Life

1. Wigglesworth, V. B. (1959). Metamorphosis, polymorphism, differentiation. *Sci Amer* 200: 100–110. See also Wigglesworth, V. B. (1979). Some memories: Interview with Sir Vincent Wigglesworth, Fellow and Emeritus Quick Professor of Biology. *The Caian*: 30–44.
2. Ibid.
3. Wigglesworth (1976): chapter 12: 138 *et seq*.
4. Ibid.
5. At this time Fabre had been translated into English, and his books were popular among the natural-history-minded English middle classes. In a subsequent interview with Wigglesworth's son, Professor Jonathan Wigglesworth, he confirmed that his father had read and was abundantly

familiar with Fabre. Indeed, Wigglesworth in turn gave a book by Fabre to his son as a child. From this interview I also gathered that V. B. Wigglesworth was also a practicing Christian, but, like most British Christian scientists, he saw no contradiction in accepting Darwin's concept of evolution.
6. Wigglesworth (1964): 1.
7. Wigglesworth (1976): 138.
8. Ibid., 139.
9. Mortimer, P. P. (2002). The control of yellow fever: A centennial account. *Microbiology Today* 29: 24–26.
10. Wigglesworth (1979): 39.
11. Edwards, J. S. (1998). Sir Vincent Wigglesworth and the coming of age of insect development. *Int J Dev Biol* 42: 471–73.

Chapter 10: Elementary Questions and Deductions

1. Wigglesworth, V. B. (1933). The physiology of the cuticle and of ecdysis in *Rhodnius prolixus* (Triatomidae. Hemiptera); with special reference to the function of the oenocytes and the dermal glands. *Quart J Micr Sci* 76: 269–318.
2. Ibid.
3. Wigglesworth (1964): 10.
4. Engel, M. S., and D. A. Grimaldi (2004). A new light shed on the oldest insect. *Nature* 427: 627–30. See also H. Muir (2004). Earliest ever insect fossil springs winged surprise. *New Sci*, 14 February: 9.

Chapter 11: The Phoenix in Its Crucible

1. Wigglesworth, V. B. (1954): 8.
2. Ibid.
3. Ibid., 5.
4. See, for example, H. C. Bennet-Clark (1979). The effect of air resistance on the jumping performance of insects. *J Exp Biol* 82: 105–21. Bennet-Clark (1992). A model of the mechanism of sound production in cicadas. *J Exp Biol* 173: 123–53. Bennet-Clark (1972). The stability of swivel wing supersonic aircraft. *Nature* 239: 451–52.
5. There is a detailed study of Lehmann in my book, *The Forgotten Plague*.
6. Interviews and communications with Henry Bennet-Clark.
7. Interview with Simon Maddrell.

Chapter 12: Two Souls in One Body

1. Wigglesworth (1933).
2. I personally favor a much commoner and more prosaic explanation of irritable bowel, which is often stress related, and perhaps complicated by diverticular disease—with perhaps his terminal illness brought about by one or more of the unpleasant long-term complications, such as stricture development or diverticular abscess and peritonitis. Another altogether commonplace and plausible explanation for his death, given his age and symptoms, might be a slow-growing colonic cancer.
3. Wigglesworth, V. B. (1934.) Factors controlling moulting and "metamorphosis" in an insect. *Nature* 131: 725–26.
4. Kopeč, S. (1922). Studies on the necessity of the brain for the inception of insect metamorphosis. *Biol Bull* 42: 322–42.

5. Kopeč, S. (1927). Über die Entwicklung der Insecten unter dem Einfluss der Vitaminzugabe. *Biologica Generalis* 3: 375–84.

Chapter 13: Bizarre Extrapolations

1. See the developmental biology online article at http://8e.devbio.com/article.php?id=9.
2. See, for example, H. L. Fevold, F. L. Hisaw, and S. L. Leonard (1931). The gonad stimulating and the luteinizing hormones of the anterior lobe of the hypophesis. *Am J Physiol* 97: 291–301.
3. Wigglesworth's subsequent recognition of the importance of Kopeč's work led to a belated recognition of the Polish scientist, who enjoyed a brief fame as a pioneer of endocrinology before his untimely death in war-torn Poland in 1943.
4. Wigglesworth, V. B. (1934b). The physiology of ecdysis in Rhodnius prolixus (Hemiptera). II. Factors controlling moulting and "metamorphosis." *Quart J Micro Sci* 77: 191–222.
5. Ibid., 197; also Wigglesworth (1959).
6. Wigglesworth (1934b): 211.
7. Wigglesworth, V. B. (1936). The function of the corpus allatum in the growth and reproduction of *Rhodnius prolixus* (Hemiptera). *Quart J Micro Sci* 79: 91–121.
8. Wigglesworth, V. B. (1940). The determination of characters at metamorphosis in *Rhodnius prolixus* (Hemiptera). *J Exp Biol* 17: 201–22.
9. Ibid: 221.
10. Bounhiol, J. J. (1938). Recherches expérimentales sur le déterminisme de la metamorphose chez le les Lépidoptères. *Bull Biol de France et de Belgique,* Suppl. 24: 1–199. Biography of Bounhiol at M. Lamy and M. Delsol (1980). Jean-Jacques Bounhiol, 1905–1979. *Ann Endocrinol* (Paris) 41: 153–56. E. Plagge (1938). Weitere Untersuchungen über das Verpuppungshormon bei Schmetterlingen. *Biol Zbl* 58: 1–12.
11. Wigglesworth (1954): 26.
12. Hachlow, V. (1931). Zur Entwicklungsmechanik der Schmetterlinge. *Roux Arch EntwMech. Organ* 125: 26–49.
13. Fukuda, S. (1940). Hormonal control of moulting and pupation in the silkworm. *Proc Imp Acad Japan* 16: 417–20. Fukuda (1941). Role of the prothroracic gland in differentiation of the imaginal characters in the silkworm pupa. *Annot Zool Japan* 20: 9–13. Fukuda (1944). The hormonal mechanism of larval molting and metamorphosis in the silkworm. *J Fac Sci Tokyo Univ* 6: 477–532.
14. Toyama, K. (1902). Contributions to the study of silk-worms. I. On the embryology of the silk-worm. *Bull Coll Agric Tokyo Imp Univ* 5: 73–118. O. Ke (1930). Morphological variation of the prothoracic gland in the domestic and wild silkworm. (In Japanese). *Bult Sci Fak Terkult, Kjusu Imp Univ* 4: 12–21.
15. Interview with Professor Chris Curtis. See also C. M. Clark and N. M. Mackintosh (1954), which contains images of the bomb damage.
16. Interview Jonathan Wigglesworth.
17. Bennet-Clark interview.

Chapter 14: Assembling the Jigsaw Puzzle

1. Williams, C. M., and A. H. Clarke (1937). Records of *Argynnis diana* and of some other butterflies from Virginia. *J Wash Acad Sci* 27: 209–13.

2. Pappenheimer, A. M. (1995). Carroll Milton Williams: December 2, 1916–October 11, 1991. In *Biographical memoirs* 5.68: 413–33. National Academy of Sciences Press.
3. Ibid.
4. Gordon, N. (1959). Juvenile hormone research reaches into fantastic realms. *Boston Sunday Herald*, August 2. Article kindly provided by Harvard Archives.
5. Cousin, G. (1932). Étude expérimentale de la diapause des insects. *Bull Biol* Suppl. 15, 1–341.
6. Williams, C. M. (1946). Continuous anesthesia for insects. *Science* 103: 57.
7. Williams, C. M. (1946b). Physiology of insect diapause: The role of the brain in the production and termination of pupal dormancy in the giant silkworm, *Platysamia cecropia. Biol Bull* 90: 234–43.
8. Ibid.
9. Ibid.
10. Williams, C. M. (1947). Physiology of insect diapause. II. Interaction between the pupal brain and prothoracic glands in the metamorphosis of the giant silkworm, *Platysamia Cecropia. Biol Bull* 93: 89–98.
11. Williams (1947).
12. Ibid. One gland was not enough—it was too difficult to dissect it out in its entirety.
13. Ibid.
14. Williams, C. M. (1948). The physiology of insect diapause. III. The prothoracic glands in the cecropia silkworm, with special reference to their significance in embryonic and postembryonic development. *Biol Bull* 94: 60–65. Williams (1952). The physiology of insect diapause. IV. The brain and prothoracic glands as an endocrine system in the cecropia silkworm. *Biol Bull* 20: 120–38. See also, Williams (1949). The prothoracic glands of insects in retrospect and in prospect. *Biol Bull* 97: 111–14.
15. Pappenheimer (1995).
16. Ibid.
17. Interview with Riddiford and Truman.
18. Maddrell, eulogy for Caian.

Chapter 15: Ecology's Magic Bullet

1. Butenandt, A., and P. Karlson (1954). Uber die isolierung eines Metamorphosehormones der Insekten in Kristallisierter Form. *Z Naturforsch*, 9b: 389–91. See also, P. Karlson, H. Hoffmester, et al. (1963). Zur chemie des ecdysons. *Liebigs Ann Chem* 662: 1–20.
2. Pappenheimer (1995).
3. Williams, C. M. (1956). The juvenile hormone of insects. *Nature* 178: 212–13.
4. McElheny, V. R. (1967). Key found to insect self-curbs. *Boston Globe*, December 27. Kindly provided by the Harvard Archives.
5. Pappenheimer (1995). See also L. M. Riddiford and C. M. Williams (1967). Volatile principle from oak leaves: Role in sex life of the polyphemus moth. *Science* 155: 589–90.
6. Williams, C. M. (1961). The juvenile hormone. I. Endocrine activity of the corpora allata of the adult Cecropia silkworm. *Biol Bull*: 116: 323–38. Williams (1961). The juvenile hormone. II. Its role in the endocrine control of molting, pupation and adult development in the cecropia silkworm. *Biol Bull*

121: 572–85. Williams (1963). The juvenile hormone. III. Its accumulation and storage in the abdomen of certain male moths. *Biol Bull* 124: 355–67. Williams and J. H. Law (1965a). The juvenile hormone. IV. Its extraction, assay, and purification. *J Insect Physiol* 11: 569–80. Williams and K. Slama K (1965b). Juvenile hormone activity for the bug *Pyrrhocoris apterus. PNAS* 54: 411–14.

7. In fact the hormone is inactive in the form it first enters the blood and has to be converted into its active form, 20-hydroxyecdysterone, in the peripheral tissues. This introduces an extra layer of complexity to its action—prior cellular activation.
8. Williams, C., and A. Spielman (1966). Lethal effects of synthetic juvenile hormone on larvae of the yellow fever mosquito, *Aedes egypti. Science* 154: 1043–44.
9. Williams, C. M. (1967). Third-generation pesticides. *Sci Am* 217: 13–17.
10. Williams, C. M., and L. Riddiford (1967). The effects of juvenile hormone analogues on the embryonic development of silkworms. *PNAS* 57: 595–601.
11. McElheny (1967). Key found to insect self-curbs.
12. Yapabandara, A. M. G. M., C. F. Curtis, et al. (2001). Control of malaria vectors with insect growth regulator pyriproxifen a gem-mining area of Sri Lanka (2001). *Acta Tropica* 80: 265–76.
13. Nijhout, H. F. (1999). Hormonal control in larval development and evolution—insects. Chapter 7 in Hall and Wake, eds. (1999).

Chapter 16: On the Steps of York Minster

1. Williamson (1992).

Chapter 17: The First Genetic Testing

1. Personal communication with Mike Hart. Richard Strathmann's research interest viewpoint is derived from his Web site at the University of Washington.
2. Strathmann, R. R. (1993). Larvae and evolution: Towards a new zoology. *Quart Rev Biol* 68: 280–82.
3. Personal communication with Mike Hart.
4. Williamson (1992): 183.
5. Personal communication with Mike Hart.
6. Hart, M. W. (1996). Testing cold fusion of phyla: Maternity in a tunicate × sea urchin hybrid determined from DNA comparisons. *Evolution* 50: 1713–18.

Chapter 18: The Evolutionary Potential of Hybridization

1. Raff, R. A. (1987). Constraint, flexibility, and phylogenetic history in the evolution of direct development in sea urchins. *Deve Biol* 119: 6–19.
2. Raff, E. C., E. M. Popodid, et al. (1999). A novel ontogenic pathway in hybrid embryos between species with different modes of development. *Development* 126: 1937–45.
3. Raff, R. A., L. Herlands et al. (1990). Evolutionary modification of echinoid sperm correlates with developmental mode. *Dev Growth Differ* 32: 283–91.
4. Rahman, M. A., T. Uehara, and J. S. Pearse (2001). Hybrids of two closely related tropical sea urchins (Genus *Echinometra*): Evidence against postzygotic isolating mechanisms. *Biol Bull* 200: 97–106.
5. Uehara, T., H. Asakura, and Y. Arakaki (1990). Fertilization blockage and

hybridisation among species of sea urchins. In M. Hoshi and O. Yamashita, eds. (1990).
6. As reference 2.
7. Willmer (1990): 117 *et seq*.
8. Personal communication, Mark Nielsen.
9. Nielsen, M. G., K. A. Wilson, et al. (2000). Novel gene expression patterns in hybrid embryos between species with different modes of development. *Evol Dev* 2: 133–44.
10. Komatsu, M., and T. Chimura (2001). Development of hybrid embryos between species in seastars and sea urchins. Absract published in *Zoological Science* 18 (suppl): 80.
11. Chimura, T., M. Komatsu, and Y. Yamazaki (2002). Larval development and genetic detection of echinoderm hybrids using RAPD-PCR technique. Abstract published in *Zoological Science* 19?: 1458.
12. Personal communication from Kaori Wakabashi.
13. I devoted two chapters to the growing field of hybryidization in my book *Virolution* (2009). More information will be found at www.fprbooks.com.

Chapter 19: A New Life-Form

1. Seb Holmes interview (2002).
2. Personal communication, Lynn Margulis.
3. Holmes interview (2002).
4. Holmes, S. P., D. W. Williamson, and N. Boetbook (2003). Phylogenetic hybridisation: a source of genetic novelty in contravention of Dollo's Law? Presented as a poster at the 38th European Marine Biology Symposium in Alverio, Portugal. See also, Williamson DI Larval transfer: Evolution by hybridisation. In L. Margulis and C. A. Asikainen, eds. (2006).
5. Ryan, F. (2009). Also Ryan F. Ryan FP. An alternative approach to medical genetics based on modern evolutionary biology. Part 5: Epigenetics and genomics. *Journal of the Royal Society of Medicine* 2009: 102: 530–37.
6. Gilbert (2003): 33.
7. Bengston, S. and Z. Yue (1997). Fossilized metazoan embryos from the earliest Cambrian. *Science* 277: 1645–48.
8. Rasmussen, B., Bengston, et al. (2002). Discoidal impressions and trace-like fossils more than 1200 million years old. *Science* 296: 1112–15.
9. Chen, Jun-Yuan, P. Oliveri, et al. (2000). Precambrian animal diversity: Putative phosphatized embryos from the Doushantuo formation of China. *PNAS* 97: 4457–62.
10. Xiao S., X. Yuan, et al. (2002). Microscopic carbonaceous compression in a terminal Proterozoic shale: A systematic reassessment of the Miaohe biota, South China. *J Paleontol* 76: 347–76.

Chapter 20: The Puzzle of the Hornworm Brain

1. Truman, J. W., and L. M. Riddiford (1999). The origin of insect metamorphosis. *Nature* 401: 447–52.
2. Riddiford, L. M., and C. M. Williams (1967). Chemical signalling between polyphemus moths and between moths and host plant. *Science* 156: 541.
3. Goodman, C. S., and B. C. Coughlin (2000). The evolution of evo-devo biology. *Proceedings of the National Academy of Sciences* 97: 4424–25.
4. Carroll, S. B. (2005). Evolution at two levels: on genes and form. *PLoS Biology* 3: e245.

5. Nielsen, C. (2000). The origin of metamorphosis. *Evolution & Development* 2 (3): 127–29.
6. Jägersten, G. (1972).
7. Kristensen, N. P. (1975). The phylogeny of hexapod "orders." A critical view of recent accounts. *Z Zool Systematic Evolutionsforschung* 13: 1–44.
8. Judson, H. F. (1979). *The eighth day of creation.* 1995 edition:194.
9. Interview of Truman and Riddiford.
10. Horodyski, F. M., L. M. Riddiford, and J. W. Truman (1989). Isolation and expression of the eclosion hormone gene from the tobacco hornworm, *Manduca sexta. PNAS* 86: 8123–27.
11. Rankin, M.A., and L. M. Riddiford (1977). The hormonal control of migratory flight in *Oncopeltus fasciatus*: The effects of corpus cardiacum, corpus allatum and starvation on migration and reproduction. *Gen Comp. Endocrin* 33: 309–21.
12. Personal communication.
13. Prokop, A. and G. M. Technau (1991) The origins of postembryonic neuroblasts in the ventral nerve cord of Drosophila melanogaster. *Development* 111: 79–88.

Chapter 21: Aristotle or Darwin?

1. Berlese, A., (1913). Intorno alle metamorfosi degli insetti. *Redia* 9: 121–36.
2. See Nowák, V. J. A. (1959): 137.
3. Ibid: 142 *et seq*; Hinton, H. E. (1948). On the origin and function of the pupal stage. *Trans Royal Entomological Soc, London* 99: 395–409; Bennet-Clark interview.
4. It is interesting to note that Darwin, and thus Wigglesworth, followed the general interpretation first proposed by the seventeenth-century Dutch naturalist, Jan Swammerdam. See Nowák (1959): 149 *et seq.*
5. Truman, J. W., and L. M. Riddiford (2002). Endocrine insights into the evolution of metamorphosis in insects. *Ann Rev Entomol* 47: 467–500.
6. Ibid., 470. See also Corbet, P. S. (1955). The immature stages of the emperor dragonfly, *Anax imperator* Leach (Odonata: Aeshnidae). *Entomol Gaz* 6: 189–97.
7. Erezyilmaz, D., L. M. Riddiford, and J. W. Truman (2006). The pupal-specific *broad* directs progressive morphogenesis in a direct developing insect. *PNAS*, in press.

Chapter 22: Cues and Common Links

1. Judson, H. F. (1979): 218.
2. Cameron, R. A., G. Mahairas, et al. (2000). A sea urchin genome project: sequence scan, virtual map, and additional resources. *PNAS* 97: 9514–18.
3. Davidson, E. H. (2001): 19.
4. Carroll, S. B. (2005). Evolution at two levels: On genes and form. *PLoS Biology* 3: 1159–66. (I should express my thanks to Alan Chalk who pointed out this article to me. It is available on open access at www.plosbiology.org)
5. Gilbert, S. F. Web reference. Developmental biology, Part 4, 21: Environmental Regulation of Normal Development. At http://www.ncbi.nlm.nih.gov/entrez/query.fcgi?cmd=Search&db=books&doptcmdl=GenBookHL&term=metamorphosis+AND+dbio%5Bbook%5D+AND+132381%5Buid%5D&rid=dbio.section.5122#5147

6. Nijhout (1994).
7. Jackson, D., et al. (2002). Ecological regulation of development: induction of marine invertebrate metamorphosis. *In J Dev* 46: 679–86.
8. For a general discussion of environmental regulation of animal development, see chapter 22 in Gilbert (2003). For a specific discussion of the bacterial symbionts in animal development see, McFall-Ngai, M. J. (2002). Unseen forces: The influence of bacteria on animal development. *Dev Biol* 242: 1–14. McFall-Ngai and E. G. Ruby (1991). Symbiont recognition and subsequent morphogenesis as early events in an animal-bacterial mutualism. *Science* 254: 1491–94. Montgomery, M. K., and
M. J. McFall-Ngai (1995). The inductive role of bacterial symbionts in the morphogenesis of a squid light organ. *Am Zool* 35: 372–80.
9. Grant-Downton, R. T., and H. G. Dickinson (2006). Epigenetics and its implications for plant biology 2. The "epigenetic epiphany": epigenetics, evolution and beyond. *Annals of Botany* 97: 11–27.
10. Beerman, W. (1952). Chromosomeren Konstanz und Spezifische Modifikation der Chromosomerenstruktur in der Entwicklung und Organ differentzierung von *Chironomus tentans*. *Chromosoma* 5: 139–384.
11. Clever, U., and P. Karlson (1960). Induktion von Puff-Veranderungen in den Speicheldrüsenchromosomen von *Chironomus tentans* durch ecdysone. *Exp Cell Res* 20: 623–26.
12. Lezzi, M. (1996). Chromosome puffing: Supramolecular aspects of ecdysone action. Chapter 4 in Gilbert, Tata, et al. (1996).
13. Cherbas, P., and L. Cherbas (1996). Molecular aspects of ecdysteroid hormone action. Chapter 5 in Gilbert, Tata, et al. (1996).
14. Gilbert, Tata, et al. (1996).
15. Ibid., xv.

Chapter 23: A Tale in a Tail

1. Barrington (1968).
2. We might also recall that there are adult forms, known as larvaceans, within the same phylum of the urochordates that are tailed tadpoles, possess gonads, and that independently feed and reproduce.
3. Barrington (1968).
4. Ibid.
5. Ibid.
6. Jackson et al. (2002).
7. Woods, R. G., K. E. Roper, M. Gauthier, et al. (2004). Gene expression during early ascidian metamorphosis requires signalling by Hemps, an EGF-like protein. *Development* 131: 2921–33.
8. Patricolo, E., M. Cammarata, and P. D'Agati (2001). Presence of thyroid hormones in ascidian larvae and their involvement in metamorphosis. *J Exp Zool* 290: 426–30.
9. Patricolo, E., C. Mansueto, P. D'Agati, and L. Pellerito (2001). Organometallic complexes with biological molecules: XVI. Endocrine disruption affects of tributyltin(IV)chloride on metamorphosis of the ascidian larva. *Applied Organometallic Chemistry* 15: 916–23.
10. Carosa, E., A. Fanelli, et al. (1998). *Ciona intestinalis* nuclear receptor 1: A member of steroid/thyroid hormone receptor family. *PNAS* 95: 11152–57; see also C. Devine, V. F. Hinman, and B. M. Degnan (2002). Evolution

and developmental expression of nuclear receptor genes in the ascidian *Herdmania. Int J Dev Biol* 46: 687–92.

11. Johnson, L. G. (1998). Stage-dependent thyroxine effects on sea urchin development. *New Zealand Journal of Marine and Freshwater Research* 32: 531–36; A. Heyland, A. M. Reitzel, and J. Hodin (2004). Thyroid hormones determine developmental mode in sand dollars (Echinodermata: Echinoidea). *Evolution and Development* 6: 382–92.
12. Youson, J. H. (1997). Is lamprey metamorphosis regulated by thyroid hormone? *Amer Zool* 37: 441–60.
13. Murr, E., and A. Sklower (1928). Untersuchungen über der inkretorischen Organe der Fische. I. Das Verhalten der schilddruse in der Metamorphose des Aales. *Z Vergl Physiol* 7: 279–88.

Chapter 24: Of Frogs and Their Relatives

1. An entertaining as well as usefully historical account of these early tetrapods is to be found in Gould (1993), chapter 4.
2. Clack, J. A. (2005). Getting a leg up on land. *Scientific American* (December 2005): 80–87; Holmes B (2006). Meet your ancestor. *New Scientist* 9 (September): 35–39.
3. Dent, J. N. (1968). Survey of amphibian metamorphosis. Chapter 7 in Etkin and Gilbert, eds. (1968)
4. Etkin and Gilbert, eds. (1968); Gilbert, Tata, et al. (1996); in particular Y. Katsutoshi, Cell death and histolysis in amphibian tail during metamorphosis. Chapter 19 in Gilbert, Tata et al. (1996).
5. Kaltenbach, J. C. (1996). Endocrinology of amphibian metamorphosis. Chapter 11 in Gilbert, Tata, et al. (1996).
6. See, for example, A. Kanamori and D. D. Brown (1996). The analysis of complex developmental programmes: Amphibian metamorphosis. *Genes to Cells* 1: 429–35.
7. See, http://www.bio.jhu.edu/Faculty/Brown/Default.html. See also, L. Cai and D. D. Brown (2004). Expression of type II iodothyronine deiodinase marks the time that a tissue responds to thyroid hormone-induced metamorphosis in *Xenopus laevis. Dev Biol* 266: 87–95.; see also N. Marsh-Armstrong, L. Cai, and D. D. Brown (2004). Thyroid hormone controls the development of connections between the spinal cord and limbs during *Xenopus laevis* metamorphosis. *PNAS* 101: 165–70.
8. Gehring, W. J. (1998).
9. Ferrier, D. E. K., and C. Minguillón (2003). Evolution of the Hox/ParaHox gene clusers. *Int J Dev Biol* 47: 605–11.

Chapter 25: The Wonder of Development

1. Bronowski (1973): 19.
2. See the URL, http://www.devbio.com/article.php?ch=2&id=9.
3. Gilbert and Epiel, eds. (2009): 221–29.
4. See http://en.wikipedia.org/wiki/Puberty.
5. See http://www.devbio.com/article.php?ch=2&id=8.
6. Mohun, T., and J. Smith (2000). Of frogs and men. Mill Hill Essays: The National Institute of Medical Research. At http://www.nimr.mrc.ac.uk/millhillessays/2000/frogs.htm.
7. Ryan (2009).

8. Ibid.
9. Ryan, F. I Virus. *New Scientist* 30 (January 2010): 32–35. See also Ryan (2009). An alternative approach to medical genetics based on modern evolutionary biology. Part 2: retroviral symbiosis. *Journal of the Royal Society of Medicine* 2009: 102: 324–31.
10. Ryan, F. P. (2006). Genomic creativity and natural selection: A modern synthesis. *Biological Journal of the Linnean Society* 88: 655–72.
11. See reference 2. Also B. Bogin (1997). Evolutionary hypotheses for human childhood. *Yrbk Physic Anthropol* 40: 63–89.
12. Rose, S. (1998). *Lifelines: Biology beyond determininsm.* Oxford: Oxford University Press.

Chapter 26: The Human Brain

1. Zoeller, T. (2004). Editorial: Local control of the timing of thyroid hormone action in the developing human brain. *J of Clinical Endocrinology & Metabolism* 89: 3114–16.
2. Kilby, C. S. Thyroid hormone and central nervous system development. *J Endocrinology* 165: 1–8; T. Vulsma, M. H. Gons, and J. de Vijlder (1989). Maternal-fed transfer of thyroxine in congenital hypothyroidism due to a total organifiction defect of thyroid agenesis. *N Eng J Med* 321: 13–16; B. Contempre, E. Jauniaux, et al. (1993). Detection of thyroid hormones in human embryonic cavities during the first trimester of pregnancy. *J Clin Endocrinol Metab* 77: 1710–22.
3. Bernal, J., and F. Pekonen (1984). Ontogenesis of the nuclear 3.5.3'-triiodothyroxine receptor in the human fetal brain. *Endocrinology* 114: 677–79.
4. Bradley, D.J., H. C. Towle, and W. S. Young (1992). Spatial and temporal expression of α- and β- thyroid hormone receptor mRNAs, including the β_2 subtype, in the developing mammalian nervous system. *J Neuroscience* 12: 2288–2302.
5. Dowling, A. L., Martz, et al. (2000). Acute changes in maternal thyroid hormone induce rapid and transient changes in gene expression in fetal rat brain. *J Neuroscience* 20: 2255–65.
6. Becker, K. B., K. C. Stephens, et al. (1997). The type 2 and type 3 iodothyronine deiodinases play important roles in coordinating development in *Rana catesbeiana* tadpoles. *Endocrinology* 138: 2989–97.
7. Marsh-Armstrong, N., H. Huant, et al. (1999). Asymmetric growth and development of the *Xenopus laevis* retina during metamorphosis is controlled by type III deiodinase. *Neuron* 24: 871–78.
8. See Zoeller in 1 above; also B. Clancy, R. B. Darlington, and B. L. Finlay (2001). Translating developmental time across mammalian species. *Neuroscience* 105: 7–17.
9. Kester, M. H., R. Martinez, et al. (2004). Iodothyronine levels in the human developing brain: major regulatory roles of iodothyronine deiodinases in different areas. *J Clinical Endocrinology & Metabolism* 89: 3117–28.

Epilogue: A Sting in the Tail

1. Locke, M. (1992). Professor Sir Vincent B. Wigglesworth: Biography and contributions to insect morphology and embryology for the 1992 distinguished international award. *Int J Insect Morphol & Embryol* 21: 101–15.

2. Locke was replying to my question in a personal communication. See also the URL: http://72.14.207.104/search?q=cache:hf76K1EiUK8J:www.cs.rice.edu/~ssiyer/minstrels/poems/798.html+Updike+biography/+V.B.+Wigglesworth+wakes+at+noon&hl=en&gl=uk&ct=clnk&cd=1&ie=UTF-8.
3. Edwards, J. S. (1998). Sir Vincent Wigglesworth and the coming of age of insect development. *Int J Dev Biol* 42: 471–73. See also Edwards (1994). In memoriam: Sir Vincent Brian Wigglesworth (1899–1994).
4. Lawrence, P. A., and M. Locke (1997). A man for our season. *Nature* 386: 757–78.
5. Akman, L., A. Yamashita, et al. (2002). Genome sequence of the endocellular obligate symbiont of tsetse flies, *Wigglesworthia glossinidia*. *Nat Genet* 32: 402–7. See also news item at the url: http://72.14.207.104/search?q=cache:yPBiOLfqH_oJ:www.genomenewsnetwork.org/articles/09_02/wiggles.shtml+Akman+L/+Genome+sequence+of+the+endocellular+obligate+symbiont+of+tsetse+flies,+Wigglesworthia+glossinidia.++Nat+Genet&hl=en&gl=uk&ct=clnk&cd=2&ie=UTF-8.
6. Kuhn (1962).
7. Ibid.
8. Gould (1993): 146.
9. See Williamson (2003), chapter 14 and fig. 14.2d.
10. Williamson, D. I., and A. L. Rice (1996). Larval evolution in the Crustacea. *Crustaceana* 69 (3): 267–87.
11. Williamson (2003).
12. Williamson, D. I. (2009). Caterpillars evolved from onychophorans by hybridogenesis. *PNAS* 106: 19901–05.
13. Syvanen, M., and C. I. Kado, eds. (1998).
14. Gould (1993): 73.
15. Gould (1989): 24.
16. The actual dating of time periods around the junction of the Proterozoic and Cambrian is variable from one authority to another. Some, for example, regard the Cambrian as dating from 570 million years ago, so the Burgess Shale is Middle Cambrian, while others regard it as 545 or 540 million years ago. The only safe measure is to qualify such designations with the actual years.
17. See illustrated Vendian URL at: http://www.ucmp.berkeley.edu/vendian/critters.html. See also a very interesting and helpful site on the Ediacaran Assemblage at: http://64.233.183.104/search?q=cache:de94t0uTGZoJ:www.peripatus.gen.nz/paleontology/Ediacara.html+Gehling+JG/++Earliest+known+echinoderm+&hl=en&ie=UTF-8
18. Gehling, J. G. (1987). Earliest known echinoderm—a new Ediacaran fossil from the Pound Subgroup of South Australia. *Alcheringa* 11: 337–45.
19. Conway, M. S. (2000). The Cambrian "explosion": Slow-fuse or megatonnage? *PNAS* 97: 4426–29.
20. Babcock, L. E., W. Zhang, and S. A. Leslie (2001). The Chengjian biota: Record of the Early Cambrian diversification of life and clues to exceptional preservation of fossils. *CSA Today*: at: http://www.gsajournals.org/gsaonline/?request=get-document&doi=10.1130%2F1052-5173(2001)011%3C0004:TCBROT%3E2.0.CO%3B2
21. Darwin (1859): chapter 9.

22. Gould (1989): 14.
23. Conway Moris (1998): 104–6; 131. D. Shu, H.-L.Luo, et al. (1999). Lower Cambrian vertebrates from South China. *Nature* 402: 42–46. See also Valentine (2004): 182 and fig 5.18.
24. Whittington, H. B. (1971). Redescription of *Marrella splendens* (Trilobitoidea) from the Burgess Shale, Middle Cambrian, British Columbia. *Geological Survey of Canada Bulletin* 209: 1–24.
25. Hughes, C. P. (1975). Redescription of *Burgessia bella* from the Middle Cambrian Burgess Shale, British Columbia. *Fossils and Strata* (Oslo) 4: 415–35. There are some beautiful pictures of this creature and many others at Stefan Bengston's pdf site: http://luna.geol.niu.edu/2000_1/fossils/text.pdf
26. Gould (1989): 138.
27. Conway Morris (1998).
28. Conway Morris. S. (1976). *Nectocaris pteryx*, a new organism from the Middle Cambrian, Burgess Shale of British Columbia. *Neues Jahrbuch für Geologie und Paläontologie* 12: 705–13.
29. Conway Morris (1998): 109–10.
30. Gould (1989): 146.
31. Ibid., 160.
32. Valentine (2004): 189 *et seq.*
33. Willmer (1990): 332.
34. The ostensible theme of the conference was "The Human Brain in the Context of Natural History: 3000 Million Years of Evolution of Sensory Systems." It was held at the Rockefeller Foundations' Study and Conference Center. This was Williamson's first public address in person since his stroke. He was assisted in getting there by Robert Sternberg and he worried about his ability to present himself. In the event, he received an enthusiastic reception. The printed report of the conference is currently in press. See S. Vickers and D. I. Williamson (2011). Larval transfer: Evolution by hybridization. In L. Margulis, C. Asikainen, and W. Krumbein, eds. *Chimera and consciousness: Evolution of the sensory self.* Cambridge, Massachusetts: The MIT Press.
35. This was subsequently published by the Free Press in the UK.
36. Valentine (2004): 179 *et seq.*
37. Williamson, D. I. (2006). Hybridization in the evolution of animal form and life-cycle. *Zool. J. Linnean Soc* 148: 585–602.
38. Conway Morris (1998): 159 *et seq.* Valentine, J. W. (1975). Adaptive strategies and the origins of grades and ground plans. *American Zoologist* 15: 391–404.
39. Williamson, D. I. (2010). Larval genome transfer: hybridogenesis in animal phylogeny. *Symbiosis?*
40. Syvanen, M. and J. Ducore (2010). Whole genome comparisons reveals a possible chimeric origin for a major metazoan assemblage. *Journal of Biological Systems* 18 (2): 261–75.

SELECTED BIBLIOGRAPHY AND FURTHER READING

Alcock, J. 1997. *In a desert garden*. Tucson, AZ: The University of Arizona Press.

Arnold, M. L. 1997. *Natural hybridisation and evolution*. Oxford: Oxford University Press.

Balfour, E. M. 1880–81. *A treatise on comparative embryology*. Vol 2. London: Macmillan & Co.

Bonner, J. T. 2002. *Lives of a biologist*. Cambridge, MA: Harvard University Press.

Boucher, D. H., ed. 1985. *The biology of mutualism*. Oxford: Oxford University Press.

Bronowski, J. 1973. *The ascent of man*. BBC.

Clark, C. M., & N. M. Mackintosh. 1954. *The school and the site: A historical memoir to celebrate the twenty-fifth anniversary of the school*. London: H. K. Lewis & Co Ltd.

Conway, M. 1998. *The crucible of creation*. Oxford: Oxford University Press.

Conway, M. S. 2003. *Life's solution: Inevitable humans in a lonely universe*. Cambridge: Cambridge University Press.

Darwin, C. 1859. *On the origin of species by means of natural selection*. London: Penguin, 1982.

———. 1871. *The descent of man*. New York: Prometheus Books, 1998.

Davidson, E. H. 2001. *Genomic regulatory systems*. New York and London: Academic Press.

Dawkins, R. 1976. *The selfish gene*. Oxford: Oxford University Press.

Etkin, W., and L. I. Gilbert, eds. 1968. *Metamorphosis: A problem in developmental biology*. New York: Appleton-Century-Crofts.

Fabre, J.-H. 1878. *The life of the caterpillar*. First published in *Souvenirs Entomologiques*. English translation by Alexander Teixeira de Mattos (1912). London and New York: Hodder and Stoughton.

Fisher, R. A. 1930. *The genetical theory of natural selection*. Oxford: Clarendon Press.

Garstang, W. 1985. *Larval forms and other zoological verses*. With a foreword by M. La Barbara (reprint). Chicago: University of Chicago Press.

Gehring, W. J. 1998. *Master control genes in development and evolution: The homeobox story*. New Haven: Yale University Press.

Gilbert, L. I., J. R. Tata, et al., eds. 1996. *Metamorphosis: Postembryonic reprogramming of gene expression in amphibian and insect cells*. New York and London: Academic Press.

Gilbert, S. F. 2003. *Developmental biology*. 7th ed. Sunderland, MA: Sinauer Associates.

Gilbert, S. F. and D. Epiel, eds. 2009. *Ecological developmental biology*. Sunderland, MA: Sinauer Associates.

Giudice, G. 1973. *Developmental biology of the sea urchin embryo*. New York and London: Academic Press.

Gould, S. J. 1989. *Wonderful life*. London: Penguin, 1991.

———. 1993. *Eight little piggies*. London: Penguin, 1994.

Haldane, J. B. S. 1932. *The causes of evolution.* London and New York: Longmans.

Hall, B. K., and M. H. Wake, eds. 1999. *The origin and evolution of larval forms.* New York and London: Academic Press.

Hoshi, M., and O. Yamashita, eds. 1990. *Advances in invertebrate reproduction.* Amsterdam: Elsevier.

Huxley, J. 1942. *Evolution: The modern synthesis.* London: Allen & Unwin.

Huxley, T. H. 1893. *Darwiniana: Collected essays II.* The Darwinian hypothesis. Place: publisher.

Hyman, L. H. 1940–49. *The invertebrates.* Vols. 1–4. New York: McGraw-Hill.

Jägersten, G. 1972. *Evolution of the Metazoan life cycle.* New York and London: Academic Press.

Judson, H. F. 1979. *The eighth day of creation.* London: Penguin, 1995.

Kuhn, T. S. 1962. *The structure of scientific revolutions.* 3rd ed. Chicago: University of Chicago Press, 1996.

Lewis, W. H., ed. 1980. *Polyploidy: Biological relevance.* New York: Plenum Press.

Lotsy, J. P. 1916. *Evolution by means of hybridisation.* The Hague: M Nijhoff.

Margulis, L. 1970. *Origin of eukaryotic cells.* New Haven: Yale University Press.

Margulis, L., and K. V. Schwartz. 2000. *Five kingdoms: An illustrated guide to the phyla of life on earth.* New York: W. H. Freeman and Co.

Margulis, L., and D. Sagan. 2002. *Acquiring genomes: A theory of the origins of species.* New York: Basic Books.

Margulis, L., and C. A. Asikainen. 2006. *Chimeras and consciousness: Evolution of sensory systems.* New York: Freeman and Co.

Mayr, E., and Provine, W. B., eds. 1980. *The evolutionary synthesis.* Cambridge, MA: Harvard University Press, 1998.

Moore, R. C., ed. 1968. *Treatise on invertebrate paleontology.* Lawrence, KA: Geological Society of America and Kansas University Press.

Morris, S. C. 1998. *The crucible of creation.* Oxford: Oxford University Press.

Nijhout, H. F. 1994. *Insect hormones.* Princeton: Princeton University Press.

Novák, V. J. A. 1959. *Insect hormones.* English trans., 1966. London: Butler & Tanner Ltd.

Ohno, S. 1970. *Evolution by gene duplication.* New York: Springer.

Popper, K. 1959. *The logic of scientific discovery.* London: Routledge, 2000.

Raff, R. A. 1996. *The shape of life.* Chicago: University of Chicago Press.

Raff, R. A., and E. C. Raff. 1987. *Development as an evolutionary process.* New York: Liss.

Ryan, F. P. 1992. *Tuberculosis: The greatest story never told.* London: Swift Publishers. Published in the U.S. as *The Forgotten Plague*, Boston: Back Bay Books, 1994.

———. 1996. *Virus X.* New York: Little Brown, 1993.

———. 2002. *Darwin's blind spot.* Boston: Houghton-Mifflin.

———. 2009. *Virolution.* London: HarperCollins.

Sapp, J. 1994. *Evolution by association: A history of symbiosis.* Oxford: Oxford University Press.

Smith, J. M., and E. Szathmáry. 1995. *The major transitions in evolution.* Oxford: Oxford University Press.

———. 1999. *The origins of life.* Oxford: Oxford University Press.

Sturtevant, A. H. 1965. *A history of genetics.* New York: Harper & Row.

Syvanen, M., and C. I. Kado, eds. 1998. *Horizontal gene transfer.* London: Chapman & Hall.

Tattersall, W. M., and E. M. Sheppard. 1934. *James Johnstone memorial volume.* Liverpool: Liverpool University Press.

Valentine, J. W. 2004. *On the origin of phyla.* Chicago: University of Chicago Press.

Villarreal, L. P. 2005. *Viruses and the evolution of life.* Place: American Society for Microbiology.

Wigglesworth, V. B. 1954. *The physiology of insect metamorphosis.* Cambridge: Cambridge University Press.

———. 1964. *The life of insects.* London: Weidenfield and Nicholson.

———. 1976. *Insects and the life of man.* London: Chapman and Hall.

Williamson, D. I. 1992. *Larvae and evolution: Towards a new zoology.* London: Chapman and Hall.

———. 2003. *The origins of larvae.* Dordrecht, London, Boston: Kluwer Academic Publishers.

Willmer, P. 1990. *Invertebrate relationships: Patterns in animal evolution.* Cambridge: Cambridge University Press.

ILLUSTRATION CREDITS

Chapter 1. Sponge crab larva, copyright Donald I. Williamson
Chapter 2. Larva of *Luidia sarsi*, copyright Donald I. Williamson
Chapter 3. Bipinnaria larva of a starfish, copyright Donald I.Williamson
Chapter 4. Pluteus larva of a brittle star, copyright Donald I. Williamson
Chapter 5. Tornaria larva of an acorn worm, copyright Donald I. Williamson
Chapter 6. Trochophore larva of a mollusk, copyright Donald I. Williamson
Chapter 7. Metamorphosis of hybrid pluteus to spheroid, copyright Donald I. Williamson
Chapter 8. Great peacock moth, adapted from *grand paon de nuit* on Wikipedia Creative Commons, http://en.wikipedia.org/wiki/File:10_grand_paon_de_nuit.jpg
Chapter 9. Malaria mosquito, adapted from Anopheles_albimanus_mosquito.jpg by James Gathany {{PD-USGov-HHS-CDC}}, http://commons.wikimedia.org/wiki/File:Anopheles_albimanus_mosquito.jpg
Chapter 10. Helicopter damselfly, copyright Barbara Strnadova, www.dova-photography.com
Chapter 11. Goliath beetle, with kind permission of Serge Mallet
Chapter 12. Kissing bug, with kind permission of Professor Chris Curtis
Chapter 13. Parabiotically conjoined insects, with kind permission of Professor Jonathan Wigglesworth
Chapter 14. Giant silkworm moth, with kind permission of Lynn M. Riddiford and James W. Truman
Chapter 15. *Culex* mosquito larva, with kind permission of Iain Petrie, www.fouragesofsand.com
Chapter 16. Nauplius larva of a barnacle, copyright Donald I. Williamson
Chapter 17. Planula larva of jellyfish, with kind permission of Antoine Morin, Biodidac image data base maintained at the University of Ottawa
Chapter 18. Hypothetical dipleurula larva, copyright Cornell University, *Introductory Biology*, http://www.biog1105-1106.org/labs/deuts/chordates.html
Chapter 19. Budding hybrid spheroid, copyright Donald I. Williamson and Sebastian Holmes
Chapter 20. Tobacco hornworm, copyright Frank Ryan and Mark Salwowski
Chapter 21. Pronymph of milkweed bug, with kind permission of James W. Truman
Chapter 22. Red abalone, with kind permission of Genevieve (Genny) Anderson
Chapter 23. Tadpole larva, of sea squirt copyright Cornell University, *Introductory Biology*, http://www.biog1105-1106.org/labs/deuts/chordates.html
Chapter 24. African clawed frog, with kind permission of Kevin Snyder, the University of Calgary
Chapter 25. Human baby in the womb, "Views of a Foetus in the Womb" (c. 1510–1512): drawing by Leonardo da Vinci {{PD}}, http://www.visi.com/~reuteler/leonardo.html
Chapter 26. Human brain, medical diagram of human brain {{PD}}, http://commons.wikimedia.org/wiki/File:Human_brain.png

INDEX

ABOUT THE AUTHOR

Scientist and author Frank Ryan, a former physician, is the author of *Virolution*; *The Forgotten Plague: How the Battle Against Tuberculosis Was Won—And Lost* (a *New York Times* Nonfiction Book of the Year); *Virus X: Tracking the New Killer Plagues*; and *Darwin's Blind Spot: Evolution Beyond Natural Selection*. Ryan is among the pioneers whose innovative ideas are changing the basic foundations of evolutionary science.

He is particularly noted for explaining the symbiotic role that viruses have played in the evolution of life in general, and human life in particular—and is a leading voice on how understanding the way viruses have contributed to human development and physiology can dramatically impact modern medicine, particularly in its treatment of AIDS and other epidemics. Ryan is a fellow of the Linnean Society of London, the Royal Society of Medicine, and the Royal College of Physicians. He is also Honorary Research Fellow at the University of Sheffield, in the United Kingdom, where he focuses on developing his evolutionary concepts and translating evolutionary science into medicine practice. He lectures frequently across Europe and in the United States.

ABOUT THE FOREWORD AUTHORS

Dorion Sagan has authored numerous books including *Death and Sex* and *Notes from the Holocene: A Brief History of the Future* and co-authored *Into the Cool: Energy Flow, Thermodynamics, and Life*. His articles have appeared in *The New York Times, The New York Times Book Review, Wired, Smithsonian, The Ecologist, Natural History,* and many other publications.

Lynn Margulis, Distinguished University Professor in the Department of Geosciences at the University of Massachusetts Amherst, has been elected to the U.S. National Academy of Sciences and the Russian Academy of Natural Sciences. She received a National Medal of Science from President Clinton in 1999. She has authored and co-authored many books, including *Luminous Fish: Tales of Science and Love; Dazzle Gradually: Reflections of the Nature of Nature; Symbiotic Planet: A New Look at Evolution*; *What Is Life?;* and *Acquiring Genomes: A Theory of the Origins of Species*. Her papers are archived at the Library of Congress.